Technology for Combating WMD Terrorism

NATO Science Series

A Series presenting the results of scientific meetings supported under the NATO Science Programme.

The Series is published by IOS Press, Amsterdam, and Kluwer Academic Publishers in conjunction with the NATO Scientific Affairs Division

Sub-Series

I. Life and Behavioural Sciences	IOS Press
II. Mathematics, Physics and Chemistry	Kluwer Academic Publishers
III. Computer and Systems Science	IOS Press
IV. Earth and Environmental Sciences	Kluwer Academic Publishers
V. Science and Technology Policy	IOS Press

The NATO Science Series continues the series of books published formerly as the NATO ASI Series.

The NATO Science Programme offers support for collaboration in civil science between scientists of countries of the Euro-Atlantic Partnership Council. The types of scientific meeting generally supported are "Advanced Study Institutes" and "Advanced Research Workshops", although other types of meeting are supported from time to time. The NATO Science Series collects together the results of these meetings. The meetings are co-organized bij scientists from NATO countries and scientists from NATO's Partner countries – countries of the CIS and Central and Eastern Europe.

Advanced Study Institutes are high-level tutorial courses offering in-depth study of latest advances in a field.
Advanced Research Workshops are expert meetings aimed at critical assessment of a field, and identification of directions for future action.

As a consequence of the restructuring of the NATO Science Programme in 1999, the NATO Science Series has been re-organised and there are currently Five Sub-series as noted above. Please consult the following web sites for information on previous volumes published in the Series, as well as details of earlier Sub-series.

http://www.nato.int/science
http://www.wkap.nl
http://www.iospress.nl
http://www.wtv-books.de/nato-pco.htm

Series II: Mathematics, Physics and Chemistry – Vol. 174

Technology for Combating WMD Terrorism

edited by

Peter J. Stopa

US Army Edgewood Chemical Biological Center,
APG MD, U.S.A.

and

Zvonko Orahovec

Ministry of Defense,
Zagreb, Republic of Croatia

Kluwer Academic Publishers

Dordrecht / Boston / London

Published in cooperation with NATO Scientific Affairs Division

Proceedings of the NATO Advanced Research Workshop on
Technology for Combating WMD Terrorism
Hunt Valley, MD, U.S.A.
19–22 November 2002

A C.I.P. Catalogue record for this book is available from the Library of Congress.

ISBN 1-4020-2682-X (PB)
ISBN 1-4020-2681-1 (HB)
ISBN 1-4020-2683-8 (e-book)

Published by Kluwer Academic Publishers,
P.O. Box 17, 3300 AA Dordrecht, The Netherlands.

Sold and distributed in North, Central and South America
by Kluwer Academic Publishers,
101 Philip Drive, Norwell, MA 02061, U.S.A.

In all other countries, sold and distributed
by Kluwer Academic Publishers,
P.O. Box 322, 3300 AH Dordrecht, The Netherlands.

Printed on acid-free paper

DEDICATION

This book is dedicated to all those that have died senselessly at the hands of terrorists.

And to my mother, Hedwig J. Stopa, who taught me the joy of writing.

TABLE OF CONTENTS

PREFACE

This book presents international perspectives on chemical and biological terrorism. It consists of a compendium of reports and commentaries that were presented at a NATO Advanced Research Workshop (ARW) entitled "Technologies for Combating WMD Terrorism". This workshop was held in Hunt Valley, Maryland in November 2002 and was held concurrently with the Joint Scientific Conference on Chemical and Biological Defense Research. The ARW was funded by the NATO Scientific Affairs Committee and was also partly supported by the Joint Science and Technology Panel for Chemical and Biological Defense of the US Department of Defense. Several corporate sponsors also provided funding for several of the attendees and presenters. The authors would like to thank everyone that contributed to this successful workshop.

ACKNOWLEDGEMENTS

The authors and the organizing committee wish to thank the NATO Scientific Affairs Committee and the Joint Science and Technology Panel for Chemical and Biological Defense for their generous support and encouragement. The organizers would also like to thank the following individuals for their participation and contributions to the NATO Advanced Research Workshop: Slavko Bokan (Croatia), Michal Bartoszcze (Poland), Rocco Cassagrande (USA), Rolf Deininger (USA), Vito Del Vecchio (USA), Christoph. Dishovsky (Bulgaria), Maria-Jose Espona (Argentina), James Genovese (USA), Filiz Hincal (Turkey), Steven Kornguth (USA), Illya Kurochkin (Russia), Alfred Madhi (Albania), Stanislaw Majcherczyk (Poland), Jiri Matousek (Czech Republic), Bozidar Stojanovic (Yugoslavia), and Milos Stojilkovic (Yugoslavia).

DISCLAIMER

INTRODUCTION

Dr. Peter J. Stopa of the US Army Edgewood Chemical Biological Center, Aberdeen Proving Ground, MD, USA, and LTC Zvonko Orahovec of the Croatian Military Academy, Laboratory for NBC Protection, NBC Training Center, Zagreb, Croatia, convened the NATO Advanced Research Workshop (ARW) entitled "Technologies for Combatting WMD Terrorism". The workshop was held during the third week of November 2002, at the Hunt Valley Inn, Hunt Valley, MD, which is approximately 20 miles north of Baltimore, MD. This ARW was held concurrently with the Joint Scientific Conference on Chemical and Biological Defense Research. The purpose of this workshop was to convene international experts in the field chemical and biological defense and discuss potential approaches that can be used to combat terrorism that arises from weapons of mass destruction (WMD), such chemical, biological, nuclear, and radiological agents.

The events of the Fall of 2001 united the world in the quest to combat terrorism in all of its forms. One of the main effects has been the re-examination of the approaches to combating terrorism associated with weapons of mass destruction (WMD). Prior doctrines were largely strategic in nature; however, the anthrax attacks in the US and the increased frequency of hoax events throughout the world have caused governments and other entities to re-baseline their programs and worry about the tactical, rather than the strategic threat posed by weapons of mass destruction.

A key piece of successful national and international defense is the integration of science, policy, and procedures. The intent of this workshop was to integrate scientists and policy-makers in their quests to combat this form of terrorism. The opening session of the research conference saw policy makers from the United States and Central Europe (Majcherczyk) present their views on what needs to be done – an over arching strategy on both the global and national level. Subsequent papers discussed potential threat agents and some of the problems that these materials could pose to detection and mitigation (Orahovec; Espona; Bokan; Bartoszcze; Matousek). Chronic and acute effects from the exposure to chemical warfare agents were also present (Stojilkovic; Dishovsky). A variety of technologies that could be used to detect and mitigate WMD materials were discussed (Kurochkin; Orahovec; Deininger; Stopa; Matousek; Cassagrande; Del Vecchio). Arguments were also presented on the need to change investment strategies in this area (Kornguth). Finally, systems integration approaches were presented that could be used as paradigms for both the medical and non-medical communities (Hincal; Genovese; Stojanovic; Madhi).

The participants felt that this workshop and the subsequent publication of the proceedings are a good first step. The opinions and discussions of this group of experts could serve as a template for future directions in the strategies in the fight against WMD terrorism.

LIST OF AUTHORS

Directors

Dr. Peter J. Stopa
US Army Edgewood Chemical Biological Center
5183 Blackhawk Road
Aberdeen Proving Ground, MD
21010-5424

LTC Zvonko Orahovec
Croatian Military Academy
Laboratory for NBC Protection
NBC Training Center
Ilica 256b
HR-10 000, Zagreb Croatia

Contributors

Dr. Slavko Bokan
Croatian Military Academy
Laboratory for NBC Protection
NBC Training Center
Ilica 256b
HR-10 000, Zagreb Croatia

Dr. Habil. Michal Bartoszcze
Center for Biological Terrorism and Countermeasures
ul. Kubelska 2
24-100 Pulawy, Poland

Prof. Rolf Deininger
Professor of Environmental Health Sciences
School of Public Health
The University of Michigan, Ann Arbor, MI 48109

Prof. Vito G. Del Vecchio
Research Director
Insitute for Molecular Biology and Medicine
University of Scranton
Scranton, PA

Prof. Christopher Dishovsky
Military Medical Academy
Department of Experimental Toxicology
3 St. G. Sofiiski str.
Sofia 1606 Bulgaria

Dr. Maria-Jose Espona
Presidencia de la Nacion Argentina
Sanchez De Bustamante
2173, Piso 17 Depto J
C.P. 1425
Capital Federal, Argentina

Mr. James Genovese
US Army Edgewood Chemical Biological Center
5183 Blackhawk Road
Aberdeen Proving Ground, MD
21010-5424

Prof. Filiz Hincal
Hacettepe University
Faculty of Pharmacy
Department of Pharmaceutical Toxicology
Ankara 06100, Turkey

Prof. Steve Kornguth
Director, Countermeasures to
 Biological and Chemical Threats
The Institute for Advanced Technology
at The University of Texas
Austin, Texas

Prof. Illya Kurochkin
Dept. Of Chemical Enzymology
Moscow State University
Moscow, Russia

COL Alfred Madhi
Ministry of Defense
 RR. « QEMAL STAFA »
 PALL. 589, SH.1, Ap.1
Tirana, Albania

Prof. Stanislaw Majcherczyk
Adviser to the Head of National Security Bureau
Office of the President of Poland
Warszawa, Poland

Prof. Jiri Matousek
Masaryk University Brno
 Research Centre of Environmental Chemistry and Engineering
Kamenice 3 CZ-625 00
Brno, Czech Republic

Dr. Bozidar Stojanovic
Institute of Architecture and Urbanism of Serbia
Bulevar Kralja Aleksandra
73/II, 1100 Belgrad, Yugoslavia

Dr. Milos Stojilkovic
National Poison Control Centre
 Military Medical Academy
Crnotrvska 17, YU-11002
Belgrade, Yugoslavia

Dr. Arthur Stuempfle
Optimetrics, Inc.
Abingdon, MD

Mr. John Walther
US Army Edgewood Chemical Biological Center
5183 Blackhawk Road
Aberdeen Proving Ground, MD
21010-5424

FOREWORD

It was in the 20th century that in the first time in history, weapons of mass destruction became available. These weapons (nuclear, radiological, biological, and chemical) are characterized by a large number of casualties and/or heavy material destruction, in case they are used.

During the 20th century, the continuous development of all kinds of these weapons was pursued, which was followed by adequate changes in doctrine, strategy, and tactics connected with their use. Because those change were caused by political, technological, and other changes in the world, it is important predict or track their course in the 21st Century.

Initially these weapons were considered exclusively as strategic weapons available to a small number of states. In time, however, weapons of mass destruction became more widespread and available. The possible situations for their use has also changed

Since most developed nations were aware of the potential aftermath of these weapons, they initiated agreements and treaties with the intention to restrict the proliferation of weapons of mass destruction and to prevent their unauthorized use in the most effective way. But despite of these efforts, it has become obvious that the 21st century has brought the potential for an even more widespread use of these weapons. This use is not restricted to nations. Terrorists, either as groups or as individuals, have shown their interest in these weapons. The most recent incidences of their use have strictly been by lone or organized terrorists.

Another possibility exists that a group that does not have these weapons can attack a nation or an entity with them. Instead, they can use the existing infrastructure in the nuclear, chemical, and biotechnology industries where sabotage can cause their deliberate release, the aftermath of which would be comparable to the actual weapon itself. These releases, on the one hand, can be catastrophic or announced. On the other hand, they can be clandestine and quiet, which would not be recognized immediately, and their use could hardly be connected with the intentional use of radiological, chemical, and biological agents.

All these facts show the great need for continuous strengthening of measures for protection from weapons of mass destruction. We must plan; we must organize; we must prepare for the possible use of these weapons.

Hence, this workshop was convened to discuss the various aspects of preparedness to WMD events. The threat was discussed – Potential technological approaches to detection, protection, and mitigation were discussed – and the integration of these parameters into a system was also explored.

Thus we wanted to share the results of these discussions with you.

PETER J. STOPA
ZVONKO ORAHOVEC

xvii

THE WMD TERRORIST THREAT

THE WMD TERRORIST THREAT

The March 1995 nerve agent attack by terrorists in the Tokyo subway system has heightened concern among government and public safety officials regarding the potential for similar incidents in the United States. Until the Tokyo attack, most of the effort toward preventing or responding to terrorist activity within transportation facilities or other areas where crowds gather was focused on hostage taking, bombing, the use of firearms, and sabotage. The Tokyo incident demonstrated the potential of a new and insidious form of terrorism, with which few in the government, public safety, or medical community were prepared to cope.

Terrorist weapons can include nuclear devices, radiological material, and chemical and biological agents. The conventional wisdom is that a nuclear weapon will be very difficult for a terrorist to acquire; however, radioactive material, chemical agents, and biological agents are relatively easy to obtain, and thus pose a greater threat. Note that both the availability and the impact of chemical and biological threat materials are high, with potentially devastating consequences.

This timeline, crafted from unclassified sources, shows the increase of actual terrorist activity involving WMD agents within the United States. This slide does not cite hoaxes, such as the letter that was alleged to contain anthrax, sent in April 1997 to B'nai B'rith headquarters in Washington, D.C., as well as the

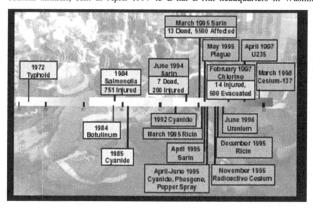

numerous hoaxes seen around the country in which letters, claiming to contain anthrax, have been sent to institutions, clinics, government buildings, and media organizations. The concern over these hoaxes is that they require the same initial response as an actual terrorist attack. The chart indicates that there was only one reported WMD event in the 1970s, with three more in the 1980s, and an exponential increase of events in the 1990s.

Some significant incidents include:

a. In 1972, members of a U.S. fascist group called Order of the Rising Sun were found in possession of 30-40 kilograms of typhoid bacteria cultures, with which they planned to contaminate water supplies in Chicago, Illinois; St. Louis, Missouri; and other large midwestern cities.

b. In 1984, two members of an Oregon cult headed by Bhagwan Shree Rajneesch cultivated salmonella (food poisoning) bacteria and used it to contaminate restaurant salad bars in an attempt to affect the outcome of a local election. Although 751 people became ill, including 45 ho spitalizations, there were no fatalities.

c. In March 1995, four members of the Minnesota Patriots Council, a right wing militia organization advocating the violent overthrow of the U.S. government, were convicted of conspiracy charges under the Biological Weapons Antiterrorism Act for planning to use ricin, a lethal biological toxin. The four men allegedly conspired to assassinate federal agents who had served papers on one of them for tax violations.

d. In May 1995, a member of the neo-Nazi organization Aryan Nations was arrested in Ohio on charges of mail and wire fraud. He allegedly misrepresented himself when ordering three vials of freeze-dried *Yersinia Pestis*, the bacteria which causes plague, from a Maryland biological laboratory.

e. In December 1995, an Arkansan was charged with possession of the toxin, ricin, in violation of the Biological Weapons Anti-terrorism Act of 1989. In 1993, Canadian customs officials had intercepted the man carrying a stack of currency with a white powder interspersed between the bills. Suspecting cocaine, customs had the material analyzed, and discovered that it was not cocaine, but ricin. He was arrested and hanged himself in his jail cell the next day.

f. In November 1995, a Chechen separatist organization left a 30-pound package o radioactive Cesium and explosives in a Moscow park. The organization informed Russian Independent Television that this was one of four such packages smuggled into Russia. Since the location of the first package was disclosed before it detonated, it is thought that the plot was to establish credibility for a possible future extortion attempt.

g. In June 1996, German authorities arrested a Slovak engineer on suspicion of smuggling 6.1 pounds of Uranium into Germany. The material was seized from a safety deposit box in Ulm, a city in southern Germany.

h. In April 1997, Russian police arrested a group that tried to sell 11 pounds of Uranium-235 stolen from a production plant in Kazakhstan. (It only takes a few pounds of enriched uranium to make a nuclear weapon.)

i. Fall 2001 – the Anthrax attacks.

As can be seen from the charts, the use of both active materials and hoax materials has been steadily increasing. Therefore, the threat is increasing and thus we must remain increasingly aware of the potential of these materials as agents of mass destruction.

Other considerations for WMD use are:

WHY WOULD TERRORISTS USE WMD?

a. Available and relatively easy to manufacture.
b. Large amounts are not needed in an enclosed space.
c. Difficult to recognize
e. Strong psychological impact;.
f. Overwhelms resources

LIMITATIONS OF WMD MATERIALS

a. Effective dissemination is difficult.
b. Delayed effects can detract from impact.
c. Counterproductive to terrorists' support.
d. Potentially hazardous to the terrorist.
e. Development and use require skill.

POTENTIAL TERRORIST TARGETS
a. Enclosed spaces.
b. Large crowds (high profile events).
c. Critical mission facilities and infrastructure.
d. Symbolic structures.
e. Accessible facilities with significant hazard/damage potential
(materials in transit).

BIOLOGICAL THREATS:
(AEROSOL, FOOD, WATER, AND UNCONVENTIONAL)

MICHAL BARTOSZCZE AND MARCIN NIEMCIEWICZ
Military Institute of Hygiene & Epidemiology
24-100 Pulawy Lubelska 2 tel/fax +48818862822
e-mail:obwwihe@obw-wihe.pulawy.pl

1. Introduction

Protection against biological weapons was previously treated very marginally in the defensive strategy of many countries. The attention was concentrated mostly on problems associated with protection against nuclear and chemical weapons. The Biological Weapons Convention, which was signed in 1972 and was about the prohibition of production, storage and use biological weapons, lay to rest the vigilance of the international community. Soon, however, it was proved that not all countries followed the rules of Convention. External to all international inspection regimes were terrorist groups that are interested in the usage of biological agents on a great scale. The Persian Gulf War changed the world opinion on these weapons with the disclosure of the Iraqi offensive biological weapons program .

Some terrorist groups also drove such programs, the Japanese Aum Shinrikyo sect being the best example. They performed experiments with anthrax and botulinum toxin, and also demonstrated interest in the bio-terrorist use of Ebola virus. Nowadays a dozen or so countries are suspected of conducting an offensive biological weapons program.

Why are biological weapons and bioterrorism a current and future threat? The attractive features of biological weapons are:
- Easy production
- High efficiency
- Difficult to detect
- Possibility of concealment
- Easy to transport
- Low price

The reasons that this type of threat is increasing are:
- Differences between poor and rich countries
- Increasing problems with fanaticism, nationalism, and intolerance
- Increasing role of terrorism as a way of achieving the terrorist groups' objectives (internal, external, and international terrorism).

The most important factor in protection against a biological weapon or a bio-terrorist attack is the ability to rapidly detect the type of biological agent being used. This allows proper prevention (prophylaxis) and medical management (leading to decreases of human losses) of casualties. Similarly the proper diagnosis and detection of the agent may also contribute to effective mitigation of the event.

2. Factors Influencing a Biological Attack

A biological attack can be executed in several different ways. These include aerosols; the use of food and water as an indirect target of attack; and an unconventional method of attack. These methods will be discussed in some detail in this section.

P. Stopa and Z. Orahovec (eds.), Technology for Combating WMD Terrorism, 5-10.
© 2004 *Kluwer Academic Publishers. Printed in the Netherlands.*

Aerosolization of the biological agent is the most effective method of attack. The following factors influence the effectiveness of a biological aerosol attack:

- Size of biological agent particles (0.1-5 µm)
- Biological factor ability to concentrate
- Biological factor resistance on the environments factors
- Optimum meteorological condition.

The aerosols themselves have the following features. They are generally invisible and odorless, making it difficult for the body to sense and attack. Similarly, since they are odorless, odor being an indicator of a vapor pressure of the material, they are difficult to detect using conventional detection approaches.

A biological attack can be executed by using long-range missiles; strategic bombers; artillery shells; vehicles (Helicopters, ships, boats, trucks, and cars) with aerosols generators; or other mobile sprayers, man-included, for bio-sabotage purposes.

The target of an attack can be strategic (a country, cities, armies etc.) or tactical (cities, civilian infrastructure (subways, electoral centers, concert halls, stadiums etc.). Similarly one may chose a target to make a political statement (assassination of national leaders or contamination of national symbols) or religious statements (assassination of religious leaders, contamination of key churches, shrines, etc.).

In the past as well in the present protection against biological aerosol attack was treated very seriously. The protection against aerosols was taken very seriously because a biological aerosol attack can cause unusually high human loss. According to data prepared for US Congress: an aerosol attack on Washington DC using 100 kg of anthrax will cause casualties comparable to an attack using a hydrogen bomb (130,000 to 3,000,000 fatalities). The US Centers for Disease Control and Prevention (CDC) estimates that the economic loss as a result of biological attack will can be as high as 26 billion US dollars.

There are several factors will influence the number and size of casualties and losses. These factors include:
- The biological agent used.
- Population susceptibility to the agent.
- The type and size of the target(s).
- The rapidity of mitigation efforts.
- Proper prophylaxis and/or treatment of casualties.

According to several sources, the highest morbidity will take place between 2 and 28 days after a biological aerosol attack (anthrax spores used for attack).

Long-range missiles or strategic bombers may also be used for a biological attack. A country that possesses an effective missile or air defense system may intercept the attacking force effectively and destroy them in time; however, not all countries possess such systems.

Nowadays we should consider the possibility of an unexpected bio-terrorist attack using unconventional means, an example of which is the newest generation of missiles, which can be launch from boat, yachts, cars, etc. This makes the protection of large municipal infrastructure difficult. Similarly, the clandestine dissemination of biological agents within facilities should also be considered. During tests conducted in the New York City subway, biological material was detected at different points throughout the system within minutes of its release. Based on this observation, it is easy to imagine the severity of an attack on the subway system used by approximately one million people every day. As of yet, there is a lack of effective solutions to this problem.

Protection against a biological aerosol attack should be based on an early warning or detection system. An effective system should give time to use in time for people to deploy personal or collective systems of protection. These detection systems are still in the developmental stage, although some countries have deployed rudimentary detection systems where appropriate. Most likely, the limiting factor of such a system will be the price (probably very high), so that the implementation of this system on a wider scale will be very difficult.

3. Aspects of Protection Against a Biological Aerosol Attack

3.1   INDIVIDUAL PROTECTION

The essential element in individual protection against biological aerosol attack is possession of an early warning system so that personnel may be given sufficient time to employ their masks. In urban environments or at high value targets, an integrated point detection system is the most likely candidate. On the battlefield, this system should be active at least a distance of 40-km from possible object of attack. This requires the use of standoff technologies, such as LIDAR, or an extensive array of point detector systems.

There is a variety of Personal Protective Equipment (PPE) that one can employ. An example is the M40 mask that is equipped with HEPA filters. These filters can stop 98% to 100% of aerosol particles in the 0.3-15 μm range. According to numerous data 26 to 51% of soldiers are not capable wearing the mask for longer than 3-4 hours. However, in temperatures of 30-40^{0}C and humidity of 50%, the time of wearing the mask is reduced to 35-60 min. This is based on experiences from the Persian Gulf War. Some soldiers also tried to sleep in their personal protective equipment, the reason being the fear of a biological attack at night, which is the most successful time to conduct such an attack. The difficulties and limitations of the widespread applications of personal protective masks include:

- The weight of the mask.
- Lifetime of the filters.
- Physiological and hygiene requirements.
- Difficulty in communication.
- Fit and maintenance of the mask.

These factors need to be considered by both military and civilian planners who are considering widespread issuance of masks, especially for protection of civilian populations. Continuous training for both the military and civilian populations in the proper fitting of the mask; exchange of filters (every 30 days in bio-safe conditions); drinking and eating; communications; and other skills.

Another component of individual protection is a personal protective suit. The personal protective suit is useful within the infected area typically for only 24 hours. The protective boots and glasses should be disinfected every 12 hours. These skills, and also proper building and equipment decontamination protocols, should be part of an effective training program for both the military and civilian populations.

3.2   COLLECTIVE PROTECTION

Collective Protection (CP) systems can be employed in the field or in large assets, such as buildings, ships, and airframes, to protect large amounts of people from the consequences of a WMD attack. They typically use some type of barrier material and a positive pressure air system. Positive pressure air eliminates the problem of possible leakage of agents into the system. A promising example is the Modular Collective Protection Equipment (MCPE). The MPCE systems are equipped with HEPA and carbon-based filters and are capable of filtering 100-200 cubic foot of air/minute. The temperature of the filtered air can also be modified and controlled. The system appears to be very useful for the demands of up to 30 people (200 – 600 cubic foot of air/minute). Impermeable shelter materials and air locks are also used in this system.

Despite the positive qualities of the system, there are some drawbacks. These include:

- Constant supply of energy (necessity for possession of a backup generator) for blowers.
- Facility of damaging the filters, thus requiring the possession of many backup filters.
- Possibility of losing the integrity of the system.

These systems are mainly in use for unit headquarters, missile and computer centers purposes, and other typically fixed sites. They are also employed in a variety of vehicles and airframes.

3.3   MEDICAL DEFENCE AGAINST BIOLOGICAL AEROSOL

The most effective methods of collective protection against biological attack are proper preventive vaccination program and proper prophylaxis. Vaccines against anthrax, botulism, tularemia, plague, Q fever, and others, are nowadays accessible. In many countries an intensive research effort for new vaccines are underway. The research uses current biotechnology principles, such as recombination technologies, so that high quality immunological and bio-safety criteria are met. According to the US Department of Defense data, the year 2003 should see the testing of many recombinant vaccines.

The points presented in this discussion shows once again that the success of all actions after a biological aerosol attack (preventive, curative, and eradicative) depends on the quick identification of the biological agent and proper management of response systems. All of these systems need to work in concert in order to assure a "bio-safe" condition for people.

4. Using Food and Water as Indirect Vectors of an Attack

4.1 FOOD

A bio-terrorist attack on a large scale may be possible using food as an indirect target or vector of the attack. Perhaps the best example of this occurred in the mid-1980's. *Salmonella typhimurium* strain was used to attack a target - meals served in luncheon bars in Dallas, Oregon. The result of this attack was that 750 persons fell ill. The intent of this attack was to influence a local election. The attack was dismissed as a natural disease outbreak until one of the persons involved the attack confessed later on. This shows that public health personnel need to be ever more vigilant in the event of an outbreak of a disease that appears to be a "natural" event.

The main biological agents that can be used in this type of attack, are the O157:H7 strains of *E. coli*, and *Shigella sp., Clostridium perfringens,* enterotoxigenic strains of *Staphylococcus aureus,* and *Clostridium* strains that produce Botulinum toxins. We should mention that the botulinum toxin shows higher stability in food in comparison to water, where the process of the hydrolysis of the toxin is considerably faster.

Contamination of food by biological agents can occur within a country or as well as outside of its borders. This can happen during manufacturing; during transport; or even during the distribution process. It is estimated that insignificant amounts of imported food are screened on a countrty's borders. Practically beyond control is genetically modified food.   The introduction of an Operation Risk Assessment system for protection against a biological attack may prove to be unusually valuable in this regard.

It is very important to test the food provided by food suppliers, especially during the large events (sports, cultural events, electoral public meetings etc.) where the possibility that a biological agent will be introduced into a very large group of people is high.

There are several steps that one can take for protection against biological agents that are introduced into food. The most important of these are stringent control of imported food and the creation of a secure system of food supervision that begins in the very early stages of production and continues through storage and distribution. There is also the necessity for creation of a laboratory system for fast detection of the biological agent in both feedstock materials as well in finished products.

4.2 WATER

Already in ancient times water was chosen as an effective element of combat with an enemy. There were well-known cases of frequent poisoning of water during retreats and seiges. In present times water can also be an indirect target of a biological attack, the possibility of which should be treated very seriously. The most likely biological agents that can be used in this type of attack are the Botulinum toxins, *V. cholerae, Bulkholderia mallei, Salmonella sp., Cryptococcus sp.* etc. Constant control of the water distribution systems, water reservoirs (storage systems) and companies that use water in their manufacturing process is the key to effective protection.

Another important element in the system is the possession of laboratories, which can provide rapid analyses the water. The recent use of bottled water by the US Army is an exakmple of such a system. Water is from a well-known source and it in constant control. The Army went to this system since disinfections tablets for water sterilization purposes were non-effective in practice. However, individual miniature systems for water filtration for soldiers in the field appear to be very promising. Additionally, the modern techniques of water sterilization, such as reverse osmosis technique, are used safely and effectively.

5. Unconventional Attack

Bio-terrorist attacks with *B. anthracis* spores in the USA showed the world the scale of a biological agent threat. Although only 22 persons were infected by *B. anthracis* spores and 5 died, antibiotic therapy was administered to 32000 people, who were largely not infected. The attack was accompanied with some panic and confusion: ransoming of medicines in drugstores; purchasing protective equipment; and changes in human behavior, such as the limitation of movement and isolation in houses. The economic loss as a result of a biological attack are very high (buildings, persons and decontamination of the environment).

Protection against an unconventional biological attack is very difficult and demands an unusual fancy for foreseeing the possible scenario of the attack. In preparation for such as attack one needs to consider the indirect attack of many target substances (cosmetics, drugs, even chewing gum). Good preparation also requires being able to anticipate the intentions and targets of the enemy through planning and exercises. Proper preparation and development of an effective defensive infrastructure will influence the outcome of protection against biological agents.

6. Impact of Genetic engineering

Development of the molecular biology methods not only resulted in improvements for the good of man, but also caused fear for using them for militarily offensive purposes. Because of genetic engineering, methods for obtaining the biological agents that are highly virulent and highly resistance on antibiotics have become possible. The efficiency of production of the pathogens and their toxins was also increased, one factor being the transfer of virulence factors from one pathogen to another. Agents obtained be these methods, a " wolf in sheep's skin ", can cause considerable difficulties in microbiological diagnostics. These modern technologies can also play a role in enhancing the pathogen's resistance to environmental factors such as temperature, light, hydrolysis, and desiccation. We should also mention about the possibility of "programming" the length of pathogen viability (life), which after "fulfilling the assignment", the pathogen can die, enabling the chance of a safe return for the winners of a conflict after extermination of enemy.

The most important aspect in the field of protection is the possession of the diagnostic capability for detection of genetically modified biological agents. The use of genetically modified biological agents is challenging for response units, which have to undertake the challenges concerning the public health aspects of an attack. Thus the basic elements of defense against bioterrorism are rapid detection, prevention of the spreading of the disease, immediate undertaking of medical assignments, and a rapid communication and information system.

In the past few years we have observed considerable progress in microbiological diagnostics, especially at the genetic level; however, such techniques are not in common use, probably due to the high cost of modern technology. Because of this, rapid reaction on the local level may not be possible in case of an attack. The limitations are also: weak laboratory infrastructure, lack of experienced staff, and impediments to laboratory standards, such as a lack of unified diagnostic procedures. We should also mention that many technologies are designed for laboratory conditions. The technologies that possess the highest value in a response should be designed and tested for field conditions. Especially essential is the continuous training of the diagnostic personnel, continuous readiness of the system, and the possession of unalterable reserves of reagents designated for identification purposes in case of bio-terrorist attack.

7. Bibliography

Bartoszcze M., Mierzejewski J.: Some problems concerning biological threats. In: Stopa P.,Bartoszcze M.Rapid Methods for Analysis of Biological Materials in the Environment. Dordecht,Boston-London: Kluwer Academic Publishers;2000,1-5.

Bartoszcze M., Niemcewicz M., Maliński M.: Recognition of biological threats. AAVDM Newsletter, 2001,6(3), p 13-14.

Bartoszcze M., Niemcewicz M., Maliński M.: Possible scenario of biological attack. AAVDM Newsletter, in press.

Bartoszcze M., Niemcewicz M., Maliński M.: The way to the Polish bio-defense system. ASA Newsletter, 2002, 02-4, p 20-21.

Bartoszcze M. NATO BW of Lab Exercise 2000. European Military Veterinary Medical Symposium ; 2000, 23-27 October; Wiesbaden,4.

Comprehensive Procedures for Collecting Environmental Samples for Culturing Bacillus anthracis. CDC Public Health Emergency Preparedness & Response. Revised April 2002, 1-12.

Croddy E.,Perez-Armendaris C.,Hart J. Chemical and Biological Warfare. Copernicus Books, New York, 2002.

Davis Christopher J.. Nuclear blindness: An overview of the Biological Weapons Programs of the Former Soviet Union and Iraq. Emerg.Inf.Dis.1999, 5,4,509-512 .

Evaluation of *Bacillus anthracis* Contamination of the Brentwood Mail Processing and Distribution C District of Columbia, October 2001.MMWR, Weekly, Dec 21, 2001, 50 (50), p 1129-1133.

Foran J.A.,Brosnan T.M. Early Warning System for Hazardous Biological Agents in Potable Water. Env.Health Persp.108,10, 2000.

Kortepeter M.G., Parker G .W. Potential Biological Weapons Threats. Emerg.Inf.Dis.1999,5(4), p. 523-527.

McCulloch S. D: Biological Warfare and the Implications of Biotechnology www.calpoy.edu/-drjones/biowar-e3.html.

Olson K.B. Aum Shinrikyo: Once and Future Threat?. Emerg.Inf.Dis.1999,5(4), p.513-516.

Pike J.: Biological Warfare Agent production www.fas.org/nuke/intro/bw/production.html.

Preparedness for the deliberate use of biological agents. A rational approach to the unthinkable. WHO/CDS/CSR/EPH/2002.16.

Russel P.K. Vaccines in Civilian Defense Against Bioterrorism. Emerg.Inf.Dis.1999, 5(4), p. 531-533.

Stopa P. Strategies for the detection of unknown biological materials. International Conference „Protection against biological threat". Warsaw, June 11, 2001, p.12.

Szczawiński J. Food as a biological weapon – need for enhanced the surveillance, international communication and cooperation. Bioterrorism –International Serveillance and Cooperation. National Security Bureau, Warszawa, 17 June, 2002.

BIOLOGICAL WARFARE AGENTS, TOXINS, VECTORS AND PESTS AS BIOLOGICAL TERRORISM AGENTS

SLAVKO BOKAN AND ZVONKO ORAHOVEC
MOD of the Republic of Croatia, Croatian Military Academy,
Laboratory for NBC Protection, HR-10000 Zagreb, Ilica 256 b, Croatia

Abstract

The threat and use of biological agents for warfare and terrorism purposes has a long history. As human, animal and plant pathogens and toxin lists will be hard to define we propose several tables of enlisted pathogens and toxins with important criteria on the basis of which a decision can be made to include in or exclude from a list of biological agents and toxins. Human, animal and plant pathogens as a biological terrorism or warfare agents have the capacity to cause disease and potentially be used to threaten humans, animals and staple crops. Human, animal and plant pathogens must be evaluated and prioritized from social-economic and significant adverse human health perspectives. This paper describes an evaluation of human, animal and plant pathogens and toxins as terrorism agents and can serve as the basis for scientific discussion or as an aid in defining the list of terrorism biological agents and toxins. This paper also shows the main vectors that can be used as a delivery system in hostile activities.

1. Introduction

Many biological agents have the capacity to cause disease and potentially be used to threaten civilian populations. The purpose of this paper is to provide information on biological agents and toxins to military and health-care providers to help them make informed decisions on protection and treatment of these agents. Although the use of biological agents and toxins in military conflicts has been a concern of military communities for many years, several recent events have increased awareness regarding the potential use of these weapons by terrorists against civilian populations.

High level of dissemination is a criterion we used in evaluation of human pathogens and toxins. The key for producing large-scale respiratory infections is to generate an aerosol of suspended microscopic droplets, each containing one to several thousands of bacterial or viral particles. The capacity for a high level of dissemination, large-scale contamination, or large area coverage as an aerosol for respiratory exposure plays the main role in the selection of a particular agent or toxin.

In the tables, the plus sign (+) signifies that the pathogen or toxin satisfies the particular criterion for inclusion in the list. The minus sign (-) signifies that the pathogen or toxin does not satisfy the criterion for inclusion in the list. In the column "Totals" you can see a number of positive answers.

According to the criterion "No effective prophylaxis or therapy", a positive answer signifies the absence of effective prophylaxis and medical treatment. The existence of immunization and appropriate treatment against a particular agent may be in inverse proportion to the likelihood that the agent will be used. There are no effective prophylaxis and therapies against the bulk of the listed agents and toxins if used as biological and toxin warfare agents since a full series of vaccinations takes at least three months and in same cases up to one year. Thus it is difficult to imagine how one might execute mass-vaccination simultaneously against more than one disease.

Toxins are effective and specific poisons produced by the metabolic activities of certain living organisms, including bacteria, insects, plants, and reptiles. They usually consist of an amino acid chain, which can vary in molecular weight between a few hundred Daltons (peptides) and one hundred thousand Daltons (proteins). They may also be low-molecular weight organic compounds. Many of them are extremely poisonous, with a toxicity that is several orders of magnitude greater than the nerve agents.

P. Stopa and Z. Orahovec (eds.), Technology for Combating WMD Terrorism, 11-28.
© 2004 *Kluwer Academic Publishers. Printed in the Netherlands.*

The research literature suggests that we have discovered the majority of the "most toxic" ($LD_{50} < 0.0025$ mg/kg) naturally occurring toxins. Because they must be delivered as respirable aerosols, their toxicities and ease of production limit a toxin's utility as an effective Mass Casualty Biological (toxin) Weapon (MCBW).

Toxins are still considered to be less suitable for dispersal on a large scale. Nonetheless, they could be used for sabotage or in special situations, e.g., against key persons. Since toxins have low volatility, they are dispersed as aerosols and then are ingested foremost through inhalation. The new micro-encapsulation technology, which is easy to use, makes it possible to protect unstable toxins when dispersed, perhaps increasing their effectiveness as biological agents.

Fifteen to twenty of some 400 known toxins have the physical characteristics that make them threats against military forces as potential MCBWs. However, many toxins could be used in weapons to produce militarily significant/terrorist (psychological) effects, especially in poorly educated civilian populations. There are still many unknowns regarding toxins and their weaponization. Some toxins can be produced by molecular biology techniques (protein toxins) or by chemical synthesis (low molecular weight toxins). A Mass Casualty Biological (toxin) Weapon (MCBW) is any toxin weapon capable of causing death or disease on a large scale, such that the military or civilian infrastructure of the state or organization being attacked is overwhelmed. A militarily significant (or terrorist) weapon is any weapon capable of affecting-directly or indirectly, physically or through psychological impact, the outcome of a military operation.

From a public health standpoint, human, animal, plant, pathogens and toxins must be evaluated and prioritized in order to assure appropriate allocation of the limited funding and resources that are often found within public health systems.

2. Criteria for Agent Selection

The criteria that are used for the evaluation of animal and plant pathogens, vectors and pests was compiled from several sources: criteria for selection of biological agents used for negotiations in Ad-hoc Group of states-parties of BTWC; the Australia Group; the US Centers for Disease Control and Prevention; Food and Agriculture Organization (FAO); and International Office of Epizootics (OIE). Rankings of animal and plant pathogens, vectors and pests as potential warfare and bioterrorism agents are shown in tables. As a result of this evaluation, the paper finally presents several lists of possible animal and plant pathogens, vectors and pests as warfare and terrorism agents. These lists prioritize and characterize the agents according to several characteristics.

2.1 CRITERIA FOR SELECTION OF HUMAN PATHOGENS AND TOXINS AS TERRORISM AGENTS

1. High level of morbidity: higher rating (++) if clinical disease requires hospitalization for treatment including supportive care and lower rating (+) if outpatient treatment is possible for most cases.

2. High level of mortality or incapacity: agents with an expected mortality of ≥50% were rated higher (+++), and with lower expected mortalities (21-49%=++, and <21%=+).

3. Stability in the environment after release (+).

4. Ease of production and transportation (+).

5. Likely methods for terrorism usage and high level of dissemination or contamination by aerosol for respiratory exposure (+++), contamination in quantities that could affect large populations (++), and dissemination potential for sabotage on food and water supply (+).

6. High potency: effective dose (ED_{50}) for biological agents or pathogens (0.1 – 15 PFU- plaque forming unit intracerebral, intraperitoneal, aerogenic or hypodermic for viruses) and (1 – 100 organisms for bacteria) (+++). For toxins: LD_{50} <0,000025 mg/kg (+++), LD_{50} from 0,000025 to 0,0025 mg/kg (++) and LD_{50} >0,0025 mg/kg (+).

7. High level of contiguousness/transmissibility or infectiousness/intoxication by variety route: direct contact (+), respiratory route (++), or both (+++).

8. No effective prophylaxis and antidotal therapy (+).

9. Special public health preparedness (+++) that requires including: stockpiling of therapeutics (+), enhanced surveillance and education (+), and improved laboratory diagnostics (+).

10. Difficult to diagnose or identify at the early stage (+).

11. Public perception: Public fear associated with an agent and the potential mass civil disruptions that may be associated with even a few cases of disease were also considered (+ to +++).

2.1.1   Human pathogens (viruses) as biological terrorism agents:

1. Lassa fever virus
2. Variola major virus (Smallpox virus)
3. Ebola virus
4. Marburg virus
5. Crimean-Congo hemorrhagic fever virus
6. Machupo virus
7. Rift Valley fever virus
8. Monkeypox virus
9. Sin Nombre virus
10. Junin virus
11. Tick-borne encephalitis virus
12. Eastern equine encephalitis virus
13. Western equine encephalitis virus
14. Venezuelan equine encephalitis virus
15. Yellow fever virus
16. Hantaan virus
17. Nipah virus
18. Chikungunya fever virus (CHIK)
19. Dengue fever virus
20. Omsk fever virus

2.1.2 Human pathogens (bacteria, rickettsiae, protozoa and fungi) as biological terrorism agents:
Bacteria/Rickettsia

1. *Bacillus anthracis*
2. *Yersinia pestis*
3. *Francisella tularensis*
4. *Rickettsia prowazekii*
5. *Rickettsia rickettsii*
6. *Bulkholderia (Pseudomonas) mallei*
7. *Bulkholderia (Pseudomonas) pseudomallei*
8. *Bulkholderia (Pseudomonas) pseudomallei*
9. *Brucella melitensis*
10. *Coxiella burnetti*
11. *Brucella abortus*
12. *Brucella suis*
13. *Chlamydia psittaci*

2.1.3 Protozoa
1. *Naegleria fowleri*
2. *Naegleria australiensis*

2.1.4 Fungi
1. *Nocardia asteroides*
2. *Coccidioides immitis*
3. *Histoplasma capsulatum*

2.2 ADDITIONAL CRITERIA FOR TOXINS

(The lower the total number means the more dangerous the toxin as warfare and terrorism agent)

1. *Toxicity*
1= Lethal dose (LD_{50}) in the 10^{-9}g/kg range.
10= Lethal dose (LD_{50}) in the 10^{-3}g/kg range.

2. *Level of incapacity or mortality*
1= Predominately cause incapacitating.
10= Predominantly lethal.

3. *Likely methods of dissemination*
1= Toxin could be aerosolized and delivered to cover large areas for aerosol contamination (large- scale dissemination). Toxin could be used in sabotage for contamination food and water.
10= Toxin could not be aerosolized and delivered to cover large areas for aerosol contamination. Toxin could be difficult used in sabotage.

4. *Stability in environment/storage*
1= Extremely stable in storage and environment.
10= Unstable in environment or requires special storage conditions.

5. *Onset*
1= Minutes or hours to onset.
10= Hours or days to onset.

6. *Ease of decontamination*
1= Extremely difficult to decontaminate after a toxin aerosol attack.
10= Decontamination would be relatively unimportant and general decontamination procedures effectively destroy toxin.

7. *Ease of production and transportation*
1= Toxin can be ease produced in large quantities - low technology, low cost, widely available (fermentation).
10= Toxin that is very difficult to produce in weaponizable quantities - high cost, only available to specialized teams (solid phase synthesis of >100 amino acid polypeptides, advanced genetic manipulation).

2.2.1 Toxins as terrorism agents:
1. Botulinum toxin
2. Ricin
3. Staphylococcal enterotoxin B (SEB)
4. Shigatoxin
5. Saxitoxin
6. Abrin
7. Trichotecene Mycotoxins (T2,DON,HT2)
8. Anatoxin A
9. Modeccin
10. Tetrodotoxin
11. Centruroides toxin
12. Toxin of Clostridium perfrigens
13. Ciguatoxin
14. Brevetoxin
15. Palytoxin
16. Cyanginosin/Microcystin
17. Batrachotoxin
18. Bungarotoxin
19. Aflatoxin
20. Viscumin
21. Viscumin
22. Volkensin

2.3 CRITERIA FOR ANIMALPATHOGENS AS BIOLOGICAL TERRORISM AGENTS

1. Agents which have severe socio-economic and/or significant adverse human health impacts (+).
2. High morbidity and/or mortality rates: agents with an expected mortality of ≥50% were rated higher (+++), and with lower expected mortalities (21-49%=++, and <21%=+).
3. Short incubation period (+).
4. High transmissibility and/or contiguousness high level of infectiousness/intoxication by contact (+), by respiratory route (++), or both (+++).
5. Low infective/toxic dose (+).
6. Difficult to diagnose/identify at an early stage (+).
7. Stability in the environment (+).
8. Lack of availability of cost effective protection / treatment (+).
9. Ease of production (+).

2.3.1 Animal pathogens as biological terrorism agents:
Viruses
1. African swine fever virus
2. Avian influenza virus (Fowl plague virus)
3. Vesicular stomatitis virus
4. Classical swine fever virus (Hog cholera virus)
5. Classical swine fever virus (Hog cholera virus)
6. Newcastle disease virus
7. Rinderpest virus
8. Pest des petits ruminants virus
9. Bluetongue virus
10. Teschen disease virus (Porcine enterovirus type 1)
11. Rift Valley fever virus
12. Nipah swine encephalitis virus
13. African horse sickness virus
14. Camel pox virus
15. Lumpy skin disease virus

2.3.2 Bacteria
1. *Bacillus anthracis*
2. *Bulkholderia (Pseudomonas) mallei*
3. *Brucella* spp.

2.3.3 Mycoplasmas
1. Contagious bovine (pleuropneum.) (*M. mycoides var. mycoides* type SC) (CBPP)
2. Contagious caprine (pleuropneum.) (*M. capriculum var. capri pneumoniae* type F38) (CCPP)

2.4 CRITERIA FOR PLANT PATHOGENS AS BIOLOGICAL TERRORISM AGENTS

1. Agents which have severe socio-economic and/or significant adverse human health impacts, due to their effect on staple crops (+).
2. Short incubation period (+).
3. Ease of dissemination: by wind (+++), by insects (++), water, etc.(+).
4. Short life cycle (+)
5. Low infective dose (+).
6. Difficult to diagnose/identify at an early stage (+).
7. Stability in the environment (+).
8. High infectivity and causes severe crop losses: ≥60% (+++), 21-59% (++), and <21%(+).
9. Lack of availability of cost effective protection/treatment (+).
10. Ease of production (+).

2.4.1 *Plant pathogens as biological terrorism agents:*
Fungi
1. *Colletotrichum coffeanum var. virulans*
2. *Puccinia graminis* (Stem Rust, Black Rust)
3. *Tilletia indica* (Carnal Bunt)
4. *Sclerotinia sclerotiorum* (Sclerotinia Stem Rot)
5. *Dothistroma pini (Scirrhia pini)* (Pine Needle Casts and Blights)
6. *Puccinia striiformis (P. glumarum)* (Stripe Rust, Yellow Rust)
7. *Pyricularia oryzae* (Rice Blast)
8. *Ustilago maydis* (Corn Smut)
9. *Claviceps purpurea* (Ergot)
10. *Peronospora hyoscyami* de Bary f.sp. *tabacina* (Adam) *skalicky* (Downy mildew)

2.4.2 Bacteria
1. Xsanthomonas albilineans (Leaf Scald)
2. Erwinia amylovora (Shoot Blight)
3. Ralstonia solanacearum (Bacterial Wilt)
4. Xsanthomonas campestris pv. citri (Citrus Cancer)
5. Xsanthomonas campestris pv. oryzae (Rice Bacterial Leaf)

2.4.3 Viruses
1. Sugar cane Fiji disease virus (Sugar cane Fiji disease)

2.5 CRITERIA FOR VECTORS OR CARRIERS OF BIOLOGICAL TERRORISM AND WARFARE AGENTS

1. Vectors known to have been produced, used or alleged to be used as weapons (+).
2. Vectors which cause significant impact on human health or animal resources (+).
3. Short life cycle (+).
4. Ease of production (+).
5. Resistance to insecticides or bio control agents (+).
6. Ease of dissemination (+).

2.5.1 *Vectors as terrorism and warfare agents:*
1. *Xenopsylla* spp.
2. *Ctenocephalis* spp.
3. *Leptopsilla* spp.
4. *Ixodides*
5. *Hyalomma marginatum*
6. *Hyalomma Anatolicum Anatolicum*
7. *Mansonia* spp.
8. *Culex* spp.
9. *Culiseta* spp.
10. *Pediculus humanus*
11. *Ixodides*
12. *Dermacentor* spp.
13. *Rhipicephalus* spp.
14. *Amblyomma* spp.
15. *Dermacentor andersoni*
16. *Dermacentor andersoni*
17. *Dermacentor varabilis*
18. *Amblyimma Cajennese*
19. *Rhipicephalus sanguineus*

2.6 CRITERIA FOR PESTS AS BIOLOGICAL TERRORISM AGENTS

1. Pests known to have been produced, used or alleged to be used as weapons (+)
2. Pests which cause sever socio-economic and/or significant adverse effect to plants (+).
3. Ease of production (+).
4. Short life cycle (+).
5. Resistance to pesticides (+).
6. High reproducibility (+).
7. Ease of dissemination (+).

2.6.1 Pests as terrorism and warfare agents:

1. *Dociostaurus maroccanus*
2. *Haplothrips Tritici*
3. *Thrips Tabaci*
4. *Eurygaster integriceps*
5. *Lygus linecularus*
6. *Acrosternum milleri*
7. *Chilo suppressalis*
8. *Cirphis unipunctata*
9. *Earias insulana*
10. *Leptinotarsa decemlineata* (Colorado potato beetle)
11. *Harmolita tritici*
12. *Phytophaya destructor*
13. *Terranychus takestani*
14. *Cenopalpus spp.*
15. *Diabrotica virgifera virgifera*

As the list of bioregulators and toxins will be hard to define for purposes of the future negotiations of the States Parties of BTWC, this paper proposes six tables of enlisted human, animal and plant pathogens and toxins with important criteria on the basis of which a decision can be made to include in or exclude from a list of the biological agents and toxins. Rankings of potential biological agents and toxins according to important criteria are shown in:

Table 1a. Assessment of Human pathogens (viruses) according to the selection criteria for biological terrorism agents.

Table 1b. Assessment of Human pathogens (bacteria, rickettsiae, protozoa and fungi) according to the selection criteria for biological terrorism agents.

Table 2a. Assessment of Toxins according to the selection criteria for biological terrorism agents.

Table 2b. Assessment of Toxin Risk (the lower the total number means the more dangerous the toxin as a warfare or terrorism agent).

Table 3. Assessment of Animal pathogens according to the selection criteria for biological terrorism agents.

Table 4. Assessment of Plant pathogens according to the selection criteria for biological terrorism agents.

Table 5. Assessment of Vectors according to the selection criteria for biological terrorism agents.

Table 6. Assessment of Pests according to the selection criteria for biological terrorism agents.

Table 1a. Assessment of Human pathogens (viruses) according to the selection criteria for biological terrorism agents.

Viruses	(1) High morbidity (++)	(2) High level of mortality/ incapacity (+++)	(3) Stability in the environment (+)	(4) Ease of production (+)	(5) High level of dissemination (+++)	(6) High potency (+++)	(7) High level of infectivity/ intoxication (+++)	(8) No effective prophylaxis and antidotal therapy (+)	(9) Special Public Health preparedness (+++)	(10) Difficulty of detection/ identification (+)	(11) Public perception (+++)	Total (24)
1. Lassa fever virus	++	++	+	+	+++	+++	+	+	+++	+	+++	22
2. Variola major virus (Smallpox virus)	+	++	+	+	+++	+++	+++	-	+++	+	+++	21
3. Ebola virus	++	+++	-	+	+++	+++	+	+	+++	+	+++	21
4. Marburg virus	++	+++	+	+	+++	+++	+	+	+++	+	+++	21
5. Crimean-Congo hemorrhagic fever virus	++	++	-	+	+++	+++	+	+	+++	+	+++	21
6. Machupo virus	++	+++	+	+	+++	+++	+	-	+++	+	+++	21
7. Rift Valley fever virus	++	++	+	+	++	+++	+	+	+++	+	+++	20
8. Monkeypox virus	+	++	+	+	+++	+++	++	+	++	+	++	19
9. Sin Nombre virus	+	+++	+	-	+++	+++	+	+	++	+	+	19
10. Junin virus	++	+++	+	+	+++	+++	+	+	++	+	++	18
11. Tick-borne encephalitis virus	++	+	+	+	+++	+++	+	-	++	+	++	17
12. Eastern equine encephalitis virus	++	+	+	+	+++	+++	+	-	++	+	++	17
13. Western equine encephalitis virus	++	+	+	+	+++	+++	+	-	++	+	++	17
14. Venezuelan equine encephalitis virus	++	+	+	+	+++	++	+	-	++	+	+	16
15. Yellow fever virus	++	++	-	+	++	++	+	-	++	+	+	14
16. Hantaan virus	+	+	-	-	++	++	++	+	none	+	+	11
17. Nipah virus	+	++	-	-	++	++	-	+	+	+	+	11
18. Chikun-Gunya fever virus (CHIK)	+	+	-	+	++	++	-	+	+	+	+	11
19. Dengue fever virus	+	+	-	+	++	++	-	+	+	+	+	11
20. Omsk fever virus	+	+	-	+	++	++	-	+	+	+	+	11

Table 1b. Assessment of Human pathogens (bacteria, rickettsiae, protozoa and fungi) according to the selection criteria for biological terrorism agents.

Disease/Pathogens: Bacteria, Rickettsiae, Protozoa, Fungi	(1) High morbidity (++)	(2) High level of mortality/incapacity (+++)	(3) Stability in the environment (+)	(4) Ease of production (+)	(5) High level of dissemination (+++)	(6) High potency (+++)	(7) High level of infectiousness/intoxication (+++)	(8) No effective prophylaxis and antidotal therapy (+)	(9) Special Public Health preparedness (+++)	(10) Difficulty of detection/identification (+)	(11) Public perception (+++)	Total (24)
Bacteria/Rickettsia												
Anthrax (Inhalational)/ *Bacillus anthracis*	++	+++	+	+	+++	+++	+++	-	+++	-	+++	22
Plague (Pneumonic)/ *Yersinia pestis*	++	+++	+	+	++	+++	+	-	++	-	+++	19
Tularemia/ *Francisella tularensis*	++	++	+	+	++	+++	+	-	++	-	++	17
Typhus exanthematicus/ *Rickettsia prowazekii*	++	+++	+	+	++	+++	+	-	++	-	++	17
Rocky Mountain Spotted Fever/ *Rickettsia rickettsii*	++	+++	+	+	+++	+++	-	+	++	+	++	19
Glanders/ *Bulkholderia (Pseudomonas) mallei*	++	+++	+	+	+++	+++	-	+	++	+	+	18
Melioidosis/ *Bulkholderia (Pseudomonas) pseudomallei*	++	+++	+	+	++	+++	-	+	++	-	+	16
Brucellosis/ *Brucella melitensis*	+	++	+	+	+++	+++	++	-	++	+	++	18
Q-fever/ *Coxiella burnetti*	++	+	+	+	+++	+++	++	-	++	+	++	18
Brucellosis/ *Brucella abortus*	+	+	+	+	+++	+++	++	-	++	-	++	16
Brucellosis/ *Brucella suis*	+	+	-	+	+++	+++	++	+	++	-	++	16
Psitaccosis/ *Chlamydia psittaci*	+	+	-	+	+++	++	++	+	++	+	+	15
Protozoa												
Primary amoebic meningoencephalitis/ *Naegleria fowleri*	+	++	-	+	+	+	+	+	++	+	+	12
Granulomatous amoebic encephalitis/ *Naegleria australiensis*	+	++	-	+	+	+	+	+	++	+	+	12
Fungi												
Nocardiosis/ *Nocardia asteroides*	++	++	+	+	+	++	+	+	++	+	+	15
Coccidioidomicosis/ *Coccidioides immitis*	+	++	+	+	+	++	+	+	++	+	+	14
Histoplasmosis/ *Histoplasma capsulatum*	++	+	+	+	+	++	+	+	++	+	+	14

Table 2a. Assessment of Toxins according to the selection criteria for biological terrorism agents.

Toxin	(1) High morbidity	(2) High level of mortality/incapacity	(3) Stability in the environment	(4) Ease of production	(5) High level of dissemination	(6) High potency	(7) High level of infectiousness/intoxication	(8) No effective prophylaxis and antidotal therapy	(9) Special Public Health preparedness	(10) Difficulty of detection/identification	(11) Public perception	Total
	(++)	(+++)	(+)	(+)	(+++)	(+++)	(+++)	(+)	(+++)	(+)	(+++)	(24)
1. Botulinum toxin	++	+++	+	+	+++	+++	+++	+	+++	+	+++	24
2. Ricin	++	+++	+	+	+++	++	++	+	+++	+	+++	22
3. Staphylococcal enterotoxin B (SEB)	++	+++	+	+	++	++	+++	+	+++	+	+++	22
4. Shigatoxin	++	+++	+	+	+	++	+++	+	+++	+	+++	21
5. Saxitoxin	++	+++	+	+	++	++	++	+	+++	+	+++	21
6. Abrin	++	++	+	+	+++	+	++	+	+++	+	+++	21
7. Trichothecene Mycotoxins (T2,DON,HT2)	++	++	+	+	+++	+	++	+	+++	+	+++	21
8. Anatoxin A	++	++	-	+	++	+	+++	+	+++	+	+++	20
9. Modeccin	++	+++	+	+	++	+	++	+	+++	+	+++	20
10. Tetrodotoxin	++	++	+	+	++	+	++	+	+++	+	+++	19
11. Centruroides toxin	++	+++	-	+	+	++	++	+	+++	+	+++	21
12. Toxin of Clostridium perfrigens	++	++	+	-	++	++	++	+	+++	+	+++	19
13. Ciguatoxin	++	++	+	-	++	++	++	+	+++	+	+++	19
14. Brevetoxin	++	++	+	-	++	+	++	+	+++	+	+++	19
15. Palytoxin	++	+++	-	-	++	+	++	+	+++	+	+++	18
16. Cyanginosin/Microcystin	++	+++	-	-	++	++	++	+	+++	+	+++	16
17. Batrachotoxin	++	++	-	-	+++	++	++	-	++	+	+++	16
18. Bungarotoxin	+	+	+	+	+++	++	+	-	++	+	+++	16
19. Aflatoxin	++	+	+	+	++	+	+	+	++	+	+++	16
20. Viscumin	++	++	+	-	++	+	+	+	++	+	+++	16
21. Verrucologen	++	++	-	-	++	+	+	+	++	+	+++	15
22. Volkensin	++	++	-	-	++	+	+	+	++	+	+++	15

Table 2b. Assessment of Toxin Risk (the lower the total number means the more dangerous the toxin as a warfare or terrorism agent).

Toxin	(1) Toxicity	(2) Onset	(3) Level of incapacity/mortality	(4) Likely methods of dissemination	(5) Stability in the environment/storage	(6) Ease of decontamination	(7) Ease of production	Total 7-70
1. Botulinum toxin	1	3	7	3	2	6	1	23
2. Shigatoxin	1	5	2	3	3	7	2	23
3. Staphylococcal enterotoxin B (SEB)	4	6	2	2	3	5	2	24
4. Trichotecene Mycotoxin (T2,DON,HT2)	7	4	7	2	1	2	2	25
5. Ricin	3	6	8	3	2	5	1	28
6. Modeccin	3	6	5	4	5	5	1	29
7. Abrin	2	6	5	5	5	5	1	29
8. Brevetoxin	6	6	2	4	2	3	8	31
9. Aflatoxin	4	8	5	5	5	1	3	31
10. Saxitoxin	3	2	8	3	3	7	5	31
11. Viscumin	3	6	5	5	6	6	1	32
12. Toxin of CL perfrigens	3	6	8	3	3	7	3	33
13. Centruroides toxin	3	4	6	5	2	5	8	33
14. Palytoxin	2	4	8	3	5	3	9	34
15. Tetrodotoxin	3	4	5	3	5	5	9	34
16. Verrucologen	3	7	6	5	6	6	3	36
17. Anatoxin A	5	1	6	7	6	8	3	36
18. Cyanginosin/Microcystin	5	2	5	3	7	7	8	37
19. Volkensin	4	5	7	6	7	5	4	38
20. Batrachotoxin	3	1	6	4	9	8	8	39
21. Bungarotoxin	3	4	6	5	8	7	8	41
22. Ciguatoxin	3	7	6	6	8	5	9	44

Table 3. Assessment of Animal Pathogens according to the selection criteria for biological terrorism agents.

Animal pathogens	(1) Severe socio-economic/ human health impacts (+)	(2) High morbidity/ mortality rates (+++)	(3) Short incubation period (+)	(4) High contagiousness/ transmissibility by contact, respiratory route, or both (+++)	(5) Low infective/ toxic dose (+)	(6) Difficult to diagnose/ identify at an early stage (+)	(7) Stability in the environment (+)	(8) Low effective or cost-effective prophylaxis/ protection/ treatment (+)	(9) Ease of production (+)	Total (13)
Viruses										
African swine fever virus	+	+++	+	+++	+	+	+	+	+	13
Avian influenza virus (Fowl plague virus)	+	+++	+	+++	+	+	+	+	+	13
Classical swine fever virus (Hog cholera v.)	+	+++	+	+++	+	+	+	+	-	12
Foot and mouth virus	+	+++	+	+++	+	+	+	-	+	12
Rinderpest virus	+	+++	+	+++	+	+	+	-	+	12
Vesicular stomatitis virus	+	+++	+	+++	+	+	+	+	+	13
Newcastle disease virus	+	+++	+	+++	+	+	+	-	+	12
Pest des petits ruminants virus	+	+++	+	+	+	+	+	-	+	10
Nipah swine encephalitis virus	+	++	+	+	+	+	+	+	-	9
Teschen disease virus (Porcine enterovirus type 1)	-	+	+	+	+	+	+	+	+	8
Camel pox virus	-	++	+	+	+	+	+	+	-	8
African horse sickness virus	+	+++	+	+	+	+	-	+	-	8
Blue tongue virus	+	+	+	+	+	+	+	-	-	7
Lumpy skin disease virus	-	+	+	+	+	+	+	+	-	7
Mycoplasmas										
Contagious bovine (pleuropneum.) (M. mycoides var. mycoides type SC) (CBPP)	-	+	-	+	+	+	+	-	+	6
Contagious caprine (pleuropneum.) (M. capriculum var. capri pneumoniae type F38) (CCPP)	-	++	-	+	+	+	+	-	+	7

Table 4. Assessment of Plant pathogens according to the selection criteria for biological terrorism agents.

Plant pathogens	(1) Severe socio-economic/ human health impacts (+)	(2) Short incubation period (+)	(3) Ease of dissemination (wind, insects, water, etc.) (+++)	(4) Short life cycle (+)	(5) Low infective dose and infectivity (+)	(6) Difficulty diagnose/ identify at an early stage (+)	(7) Stability in the environment (+)	(8) Yield loss (+++)	(9) Cost-effective protection/ treatment (+)	(10) Ease of production (+)	Total (14)
Fungi											
Colletotrichum coffeanum var. virulans	+	+	+++	+	+	+	+	+++	+	+	14
Tilletia indica	+	+	+++	+	+	+	+	++	-	+	13
Puccinia graminis	+	+	+++	+	+	+	+	++	-	+	13
Sclerotinia sclerotiorum	+	+	+++	+	+	+	+	+++	-	+	13
Puccinia striiformis (P. glumarum)	-	+	++	+	+	+	+	+	-	+	9
Pyricularia oryzae	-	+	++	-	+	+	+	++	-	+	8
Ustilago maydis	-	+	++	-	+	+	+	++	-	+	9
Dothistroma pini (Scirrhia pini)	-	+	++	-	+	+	+	++	-	+	8
Claviceps purpurea	+	+	+++	-	-	+	-	++	-	-	8
Peronospora hyoscyami de Bary f.sp. tabacina (Adam) skalicky	-	+	+++	-	-	+	+	+	-	-	7
Bacteria											
Xsanthomonas albilineans	+	+	+++	+	+	+	+	+++	+	+	14
Erwinia amylovora	+	+	+++	+	+	+	-	+++	+	+	12
Ralstonia solanacearum	+	+	++	+	+	+	+	++	-	-	11
Xsanthomonas campestris pv. citri	-	+	++	-	+	+	+	++	+	-	10
Xsanthomonas campestris pv. oryzae	-	+	++	-	+	+	-	++	+	+	9
Viruses											
Sugar cane Fiji disease virus	+	+	++	-	+	+	-	++	-	-	8

Table 5. Assessment of Vectors according to the selection criteria for biological terrorism agents.

Vectors	Order	Class	Biological Agent	Disease	(1) Weapo-nized	(2) Significant impact on human health or animal resources	(3) Short life cycle	(4) Ease of produc-tion	(5) Resistance to insecticides or bio control agents	(6) Ease of dissemi-nation	Total
					(+)	(+)	(+)	(+)	(+)	(+)	(6)
Xenopsylla spp. Ctenocephalis spp. Leptopsilla spp.	Siphonoptera	Insecta	Yersinia pestis	Plague	+	+	+	+	+	+	6
Ixodides Hyalomma marginatum Hyalomma Anatolicum Anatolicum	Acari	Arachnida	Arbovirus	Crimean-Congo hamorhagic fever (CHF)	+	+	+	+	+	+	6
Mansonia spp. Culex spp. Culiseta spp.	Diptera	Insecta	Arbovirus	Easzern Equine Encephalitis	+	+	+	+	+	+	6
Pediculus humanus	Anopluraa	Insect	Rickettsia prowasekii	Typhus exanthematicus	+	+	+	+	+	+	6
Ixodides Dermacentor spp. Rhipicephalus spp. Amblyomma spp.	Acari	Arachnida	Francisella tularensis	Tularemia	+	+	+	+	+	+	6
Dermacentor andersoni	Acari	Arachnida	Coxiella burnetti	Q-Fever	-	+	+	+	+	+	5
Dermacentor andersoni Dermacentor varabilis Amblyimma Cajennese Rhipicephalus sanguineus	Acari	Arachnida	Rickettsia rickettsii	Rocky Mountain Spoted Fever	-	+	+	+	+	+	5

Table 6. Assessment of Pests according to the selection criteria for biological terrorism agents.

Pests	Order	Common Names	Host	(1) Weaponized (+)	(2) Severe socio-economic/significant adverse effect to plants (+)	(3) Ease of production (+)	(4) Short life cycle (+)	(5) Resistance to pesticides (+)	(6) High reproducibility (+)	(7) Ease of dissemination (+)	(7) Total (7)
Dociostaurus maroccanus	Orthoptera	Grasshoppers Crickets Cockroaches	Plants	-	+	+	+	+	+	+	6
- Haplothrips Tritici - Thrips Tabaci	Thysanoptera	Thrips	- Wheat, mais - Tabacco, tomato	-	+	+	+	+	+	+	6
- Eurygaster integriceps - Lygus lineacularus - Acrosternum milleri	Hemiptera	Bugs	- Wheat - Pistachio - Pistachio	-	+	+	+	+	+	+	6
- Chilo suppressalis - Cirphis unipunctata - Earias insulana	Lepidoptera	Batterflies Moths Skippers	- Rice - Rice, mais - Cotton	-	+	+	+	+	+	+	6
Leptinotarsa decemlineata (Colorado potato beetle)	Coleoptera	Beetles Weevils	Potatoes	-	+	+	+	+	+	+	6
Harmolita tritici	Hymenoptera	Ants Bees Wasps	Wheat	-	+	+	+	+	+	+	6
Phytophaya destructor	Diptera	Flies	Wheat (Barleiy) Oats	-	+	+	+	+	+	+	6
Terranychus takestani	Tetranychidae	Mites	Plants	-	+	+	+	+	+	+	6
Cenopalpus spp.	Errophyoidae	Mites	Fruit trees	-	+	+	+	+	+	+	6
Diabrotica virgifera virgifera	Chrysomelidae	Western corn rootworm	Maize	-	+	+	+	+	+	+	6

Potential biological agents and toxins with an expected mortality of $\geq$50% were rated higher (+++) than agents with lower expected mortalities (21-49% = ++, and <21% = +).

Biological agents and toxins were rated higher for morbidity are if clinical disease required hospitalization for treatment (including supportive care) (++) , and ratted lower (+) if outpatient treatment was possible for most cases.

Biological agents and toxins were rated on their dissemination potential based on the likelihood of contamination of a large area by aerosol for respiratory exposure (+++); contamination in quantities that could affect large populations (++); and sabotage of food and water supply (+).

High level of infectivity or intoxication by variety route is showed according of the kind of exposure: direct contact (+), respiratory route (++), or both (+++).

Biological agents and toxins also were ranked based on any special public health preparedness that was required including: stockpiling of therapeutics (+), enhanced surveillance and education (+), and improved laboratory diagnostics (+).

Public fear associated with an agent and the potential mass civil disruptions that may be associated with even a few cases of disease were also considered (+ to +++).

The additional list of criteria for toxins and table of toxins should be used for a more detailed risk assessment and for comparison of the toxins on the list in which the lower the total number, the more potential the toxin has as a terrorism or warfare agent. This was done with the purpose of making the final decision on the list of toxins easier. In the additional list of criteria for toxins, one may see, at first glance, an unimportant criterion: Ease of decontamination. For toxins, decontamination would be relatively unimportant but fungal toxins are extremely difficult to decontaminate and once an area is contaminated (infected), it can take years of hard cleaning to get rid of the toxins, especially if delivered with fungal spores.

For the evaluation of toxins, the toxicity should be interpreted with caution and certainly should include agents that are non-lethal but do have military and terrorism utility. Lethality alone is not an appropriate criterion on which to base a toxin's potential. For evaluation of toxins it was preferred toxins that cause primarily incapacitating to lethal toxins because these toxins have high potency and they represent a significant threat in the future. For example Staphylococcal enterotoxin B (SEB), so-called super antigen, is one of the most potent agents for incapacitation. It can cause illness at extremely low doses, but relatively high doses are required to kill.

3. Conclusions

The tables show that there are only a few agents that can truly threaten civilian populations on a large scale. If released upon a civilian population, these agents would pose the most significant challenge for public health and medical responses. The above criteria for ranking potential biological agents could be used for the prioritization of biological threat agents for national preparedness efforts to combat bioterrorism. Having a defined method for evaluating biological threat agents allows for a more objective evaluation of newly emerging potential threat agents, as well as continued re-evaluation of established threat agents. Using this prioritization method can help focus public health activities related to bioterrorism detection and response and assist with the allocation of limited public health resources.

Many animal and plant pathogens, vectors and pests can be used as terrorism and warfare biological agents and cause illness. Transmissible animal diseases have the potential for very serious and rapid spread, irrespective of national borders. These are of serious socio-economic or public health consequence and are of major importance in the international trade of animals and animal products. Having a defined method for evaluating biological threat agents such as animal and plant pathogens, vectors and pests, allows for more objective evaluation of newly emerging potential threat agents. This method of evaluation can help focus both public health activities and agriculture activities related to bioterrorism detection and response.

4. References

1. Rotz, D. L., Khan, S. A., Lillibridge, R. S., Ostroff, M. S. and Hughes, M. J., Priotritizing potential biological terrorism agents for public health preparedness in the united states: Overview of evaluation process and identified agents, National Center for Infectious Diseases, Centers for Disease Control and Prevention, Proceedings of CBMTS III, 2000.

2. Hamilton M., Toxins and Mid-Spectrum Agents. ASA Newsletter, No. 66, 89-3, (1998).

3. Clark, K., The Chemical Weapons Convention: Chemical and Toxin Warfare Agents and Disarmament, Cranfield University, Royal Military College of Science. (1997).

4. Geissler, E., Biological and toxin weapons today. SIPRI, Oxford University Press, Oxford, (1986).

5. Daly, J. W., The chemistry of poisons in amphibian skin. Proc. Nat'l. Acad. Sci. USA. 92: 9-18 (1995).

6. FM 3-9, Chemical Agents and their properties, Headquarters Department of the Army, (1994) Washington, DC., (1994).

7. Mandell, G., Douglas, R., Bennett, J. (1990) Principles and Practice of Infectious Diseases, 3rd Edition. Churchill Livingstone, New York.

8. Murphy, B.R. and Chanock, R.M.(1985) Immunization against viruses. In: Fields, B.N.

9. Hahn, C.S., Lustig S., Strauss E.G. and J.H. Strauss. (1988) Western equine encephalitis virus is a recombinant virus. Proc. Natl. Acad. Sci. USA. 85: 5997-6001.

10. Pfaff, E., Kuhn, C., Schaller, H., Leban, J. (1985) Structural analysis of the foot-and-mouth disease virus antigenic determinant.

11. Sakaguchi, G. (1994) Clostridium botulinum toxins. Pharmacology and Therapeutics, 19: 165.

12. Morse, S.S. (ed.) (1993) Emerging Viruses. Oxford University Press.

13. Geissler, E. and Woodall J.P. (1994) Control of Dual-Threat Agents: The Vaccines for Peace Programme, SIPRI, Oxford University Press.

14. Culliton, B.J. (1990) Emerging Viruses, Emerging Threat. Science 247: 279- 280.

15. Williams, L. and Westinf, A.H. (1983) "Yellow rain" and the new threat of chemical warfare. Ambio, Stockholm.

16. Monath, T.P. (1994) Yellow fever and dengue: The interactions of virus, vector, and host in the re-emergence of epidemic disease. Semin Virol 5: 133-145.

17. C.J. Peters et al. (1994) "Filoviruses as Emerging Pathogens", Seminars in Virology, vol. 5, pp 147-154.

18. Murphy, F.A. (1994) Infectious Diseases. Adv.Vir. Res. 43: 2-52.

19. Peters, C.J. (1994) Molecular Techniques Identify a New Strain of Hantavirus. ASM News. 60, 5: 242-3.

20. Hall, S., Striclartz, G. (1990) Marine Toxins. Origin, Structure and Molecular Pharmacology. Published by Am. Chem. Society.

21. Hunter, S. (1991) Tropical Medicine. 7th edn. W.B. Saunders Company.

22. Feldmann et al. (1993) "Molecular biology and evolution of filoviruses", Arch.Virol (supp), vol. 7, pp 81-100.

23. Halstead, S.B., Hoeprich, P.D., Jordan, M.C., Ronald A.R. (1994) Infectious diseases: A treatise of infectious processes, 919-923.

24. Advisory Commission on Dangerous Pathogens (1995) Categorization of biological agents according to hazard and categories of containment, Fourth Edition 1995, Her Majesty's Stationery Office, London.

25. Kuno, G. (1995) Review of the factors modulating dengue transmission. Epidemiological Reviews, 17(2): 321-335.

26. Hughes J.M. and La Montagne J. R. (1994) The Challenges Posed by Emerging Infectious Diseases. ASM News 60, 5: 248-50.

27. C.J. Peters et al. (1991) "Filoviruses", Chapt.15 in Emerging Viruses (ed. by S. Morse, Oxford University Press, New York), pp 159-75.

28. Calnek, B.W., et al., (eds). 1997. Diseases of Poultry, 10th Ed. Iowa State University Press, Iowa. pp 583-600.

29. Scudamore J. M. 1993. Contagious Bovine Pleuropneumonia. State Veterinary Journal 3(3): 7-10. OIE (1996) World Animal Health in 1995.

30. Radostits O. M., Blood D. C., Gay C. C. 1994. Veterinary Medicine: A Textbook of the Diseases of Cattle, Sheep, Pigs, Goats and Horses. Eighth edition. Bailliere Tindall.

31. Gyles C.L., Thoen C.O. (eds). 1993. Pathogenesis of Bacterial Infections in Animals. Second Edition. Iowa State University Press / Ames.

32. Wileman T., Rouiller I. & Cobbold C. 1997. Assembly of African swine fever virus. Report, Annual Meeting of the National Swine Fever laboratories, Vienna, 55.

33. Brunt, A., Crabtree, K., Dallwitz, M., Gibbs, A. and Watson, L. 1996. Viruses of Plants: Descriptions and Lists from the VIDE Database. 1484 pp. C.A.B. International, U.K

AN EVALUATION OF BIOREGULATORS/MODULATORS AS TERRORISM AND WARFARE AGENTS

SLAVKO BOKAN AND ZVONKO ORAHOVEC

MOD of the Republic of Croatia, Croatian Military Academy,
Laboratory for NBC Protection, HR-10000 Zagreb, Ilica 256 b, Croatia

Abstract

Bioregulators or modulators are biochemical compounds, such as peptides, that occur naturally in organisms and regulate processes on the cellular or higher level. Advances in biotechnology has created the potential for the misuse of peptide bioregulators in offensive biological weapons programs. They are a new class of weapons that can damage the nervous system, alter moods, trigger psychological changes, and kill. Within neuroscience over the last twenty years has been an explosion of knowledge about the receptor systems on nerve cells that are of critical importance in receiving the chemical transmitter substances released by other nerve cells. The potential military or terrorism use of bioregulators is similar to that of toxins. Together with increased research into toxins, the bioregulators have also been studied and synthesized. This paper presents an evaluation of bioregulators that can be used in hostile activities.

1. Introduction

Biomedical science and the pharmaceutical industry are in the midst of a revolution in the science and technology of drug discovery that will significantly complicate the control of chemical and biological weapons. Bioregulators/Modulators/Regulatory Peptides/Molecular agents play one of the main roles in the Non-Lethal Chemical and Biological Weapons Program. Neuropharmacology is one of the areas in rapid expansion; the toll of mental illness, and the growing promise of chemotherapeutical treatments, makes it certain that a wide range of new psychoactive compounds will be discovered. In the near future, agents will be developed that affect perception, sensation, cognition, emotion, mood, volition, bodily control, and alertness. All bioregulators discussed here and practically all the types of potential non-lethal agents are analogues of their naturally-occurring biochemicals, because their physiological activity depends on their ability to bind at the same sites as the natural biochemicals do.

In fact, a categorical distinction between lethal and non-lethal agents is not scientifically feasible. Not only are certain individuals more susceptible to some agents, but there is also a synergy between two different non-lethal agents that may make their combination highly lethal to everyone. Rational strategies to discover such synergistic pairs will soon be available. Very large number of new, highly toxic compounds with precisely understood and controllable physiological effects will soon be discovered. Most of them will be synthesized from precursors that are not currently regulated under the CWC.

Many biological agents have the capacity to cause disease and potentially be used to threaten civilian populations. The purpose of this paper is to provide information on bioregulators to military and health-care providers at all levels to help them make informed decisions on treatments and protection from these agents.
Bioregulators can act as neurotransmitters and modify neural response. Bioregulators are closely related to substances normally found in the body that regulate normal biological processes. Some examples of potential application of bioregulators are to cause pain, as an anesthetic, and to influence blood pressure.

These substances can also modified synthetically, whereupon they may obtain new properties. It is feasible to produce some of these compounds by chemical synthesis. It is apparent that the past decade has brought an enormous increase in knowledge about the pharmacology and structural biology of receptors. In the last ten years considerable advances have taken place in this *in vitro* synthesis of

P. Stopa and Z. Orahovec (eds.), Technology for Combating WMD Terrorism, 29-40.

peptides and already commercial production in large quantities of various pharmaceutical peptides are freely available. Synthetic derivatives or slightly modified forms of these compounds can have drastically altered toxic effects and these could be important in the development of new agents. Advances in discovery of novel bioregulators, especially bioregulators for incapacitating, understanding of their mode of operation and synthetic routes for manufacture have been very rapid in recent time. Some of these compounds may be potent enough to be hundreds of times more effective than the traditional chemical warfare agents. Some very important characteristics of new bioregulators that would offer significant military advantages are novel sites of toxic action; rapid and specific effects; penetration of protective filters and equipment and militarily effective physical incapacitation.

Peptide bioregulators are interesting regulatory molecules for many reasons. Their range of activity covers the entire living system, from mental processes (e.g. endorphins) to many aspects of health such as control of mood, consciousness, temperature control, heart rate, immune responses, sleep, or emotions, exerting regulatory effects on the body. As such, they are produced in very small quantities that are essential for the normal homeostatic functions of the body. They are also capable of regulating a wide range of physiological activities, such as bronchial and vascular tone and muscle contraction. This opens an unprecedented possibility to use toxic substances that could not be traced in the human body. In each case a clandestine application of such substances can lead to death - "killing without a trace".

Although the use of bioregulators in military conflicts has been a concern of military communities for many years, several recent events have increased the awareness regarding the potential use of these weapons by terrorists against civilian populations. Although no known illnesses resulted from these attempts to use these agents as weapons, the exploration of the use of these agents by these and other extremist organizations caused great concern regarding the potential vulnerabilities of civilians to such weapons.

Bioregulators are still considered to be less suitable for dispersal on a large scale. Nonetheless, they could be used for sabotage or in especially designed situations, e.g., against key persons. Since bioregulators have low volatility, they are dispersed as aerosols and then taken up foremost through inhalation. During recent years, discussions have started on the risk of bioregulators being used as chemical warfare agents. These types of substances do not belong to the group of toxins but are, nonetheless, grouped with them since their possible use is similar.

There are still many unknowns regarding bioregulators and their weaponization. Weaponization is defined as preparing and treating a biological agent or toxin to enhance its effectiveness as a weapon, and/or inserting a biological agent into a delivery system suitable for hostile use. A Mass Casualty Biological (Toxin) Weapon (MCBTW) is any such weapon capable of causing death or disease on a large scale, such that the military or civilian infrastructure of the state or organization being attacked is overwhelmed. A militarily significant (or terrorist) weapon is any weapon capable of affecting-directly or indirectly, physically or through psychological impact, the outcome of a military operation. Advances in the use of viral and bacterial vectors enhance the possibility for direct delivery of toxin or bioregulator to the human target or they could be used to transfer the toxin or bioregulator genes to the target.

2. Some Examples of Bioregulators

Unlike conventional chemical and biological agents, most of the bioregulator compounds are not widely known. It one needed to classify them, they would be mid-spectrum between the classical chemical and biological agents.
As with classical agents, these agents can be grouped into classes. A brief description of each of the classes follows.

2.1. Endorphins (α, β, and γ-Endorphin)

Endorphins are small-chain peptides, which activate opiate receptors, producing feelings of well-being, tolerance to pain, etc. These compounds are hundreds or even thousands of times more potent than morphine on a molar basis. Because of this potency, their concentrations in vivo are low, and it has taken recent advances in experimental neuroscience to elucidate the chemistries of these hormones. The term opioid peptides are used for the endorphins. Proopiomelanocortin - POMC (pro-ACTH-Endorphin) is a glycosylated 31 kDa protein precursor posttranslational processing of which yields several neuroactive peptides upon specific cleavage and possibly a great number of as yet unidentified small peptides that may be pharmacologically active. Endorphins can further decompose into small fragments (oligomers) which are still active, and which pass the blood-brain barrier more readily. Their high activity and specificity make endorphins attractive compounds from a clinical view, but most are active only if injected into the blood (or the cerebrospinal fluid). This is because peptides are digested in the stomach, decomposed by proteolytic and other enzymes. Also, because of their size and structure, they have difficulty passing into the brain. Thus, despite the low oral to parenteral ratio of many morphine derivatives, they will probably not be replaced by small-chain peptides anytime soon. Dipeptidyl carboxypeptidase, enkephalinases, angiotensinases, and other enzymes accomplish enzymatic degradation of small-chain endorphins.

POMC cleavage products include a large N-terminal fragment, which yields γ-MSH (melanocyte stimulating hormone-gamma) and possibly other unidentified cleavage products, ACTH (corticotropin, 39 amino acids), Lipotropin, α-MSH (melanocyte stimulating hormone-alpha; melanotropin), β-MSH (melanocyte stimulating hormone-beta) and β-endorphin. Individual products of the POMC protein act on and may be produced by immune cells, thus establishing close links between immune cells and the nervous system. Endorphin molecules have a separate nomenclature (α, β, γ) that denotes their stereochemistry.

β-endorphin (and also α-endorphin and γ-endorphin derived from it) has been found to be produced also by macrophages and lymphocytes. β-endorphin appears to act differentially: its C-terminal moiety enhances T-cell proliferation, whereas this stimulatory effect can be prevented by peptides that possess the N-terminal enkephalin sequence. Human β-endorphin is the most potent of three stereoscopic variants, and has the same sequence as the C-terminal end of β-lipotropin.

Endorphins enhance the natural cytotoxicity of lymphocytes and macrophages towards tumor cells, stimulate human peripheral blood mononuclear cell chemotaxis and inhibit production of T-cell chemotactic factors. Opiate receptors presynaptically inhibit transmission of excitatory pathways including acetylcholine, the catecholamines, serotonin, and substance P, a neuropeptide active in pain neurons. Endorphins may also be involved in glucose regulation.

2.2. Substance P (SP)

Substance P, (P=powder), the designation of which originates from early studies using powdered extracts of equine brain and intestines. It is known also as neurokinin-1 (NK1). It is a member of a family of proteins known as tachykinins. This neuropeptide was found in the gut as well as in the brain. It is responsible for a number of excitatory effects on both central and peripheral neurons. It contracts smooth muscle, constricts bronchioles and increases capillary permeability. When released from afferent nerves, it causes a neurogenic inflammatory response, including mast cell degranulation. Substance P, a polypeptide (molecular weight = 1,350 D), is active in doses of less than one microgram. Substance P causes a rapid loss of blood pressure, which may cause unconsciousness.

2.3. Endothelins (ET-1, -2, -3) or (EDCF-endothelium-derived contracting factor)

Endothelins are a family of closely related peptides of 21 amino acids with two disulfide bonds. The four known species are isoforms encoded by four different genes. They are called ET-1 (endothelin-1), ET-2 (endothelin-2), ET-3 (endothelin-3) and VIC (vasoactive intestinal contractor).

Endothelin is a highly potent vasoconstrictor peptide first isolated from porcine endothelial cell supernatant. Varying amounts of ET are also produced in other cell types such as smooth muscle, neuron, mesangium, melanocyte, parathyroid and amnion. Individual ET may posses separate physiological or pathophysiological roles in different target tissues. Secretion of ET is stimulated by epinephrine, angiotensin II, arginine vasopressin, transforming growth factor beta, thrombin, interleukin-1, and hypoxia. Endothelins act to stimulate contraction of many smooth muscle tissues including blood vessels, uterus, bladder, and intestine. ET-1 is the most potent vasoconstrictor peptide yet discovered.

Numerous studies have implicated the endothelins in cardiovascular diseases such as hypertension, heart failure, and atherosclerosis. Endothelin levels are elevated in atherosclerosis, congestive cardiac failure, and renal insufficiency. ET may play an important role in homeostatic hemodynamic balance. Endogenous endothelins and ET receptor subtypes are present in various endocrine organs. ET appears to act as a modulator of secretion of prolactin, gonadotropins, GH and TSH. It is also may act as a neurotransmitter.

Among this family of peptides ET-1 is the most studied compound. Therapeutic potential of endothelins has generated tremendous interest in numerous laboratories around the world.

Structures of recently isolated snake venom sarafotoxins (**Sarafotoxin S6a** and **S6b**) bear striking resemblance to endothelins. They have vasoconstrictor activity and potent coronary constrictor activity. They can cause heart arrest in several minutes with concentration of LD_{50} - 15 mg.kg^{-1}.

2.4. Bradykinin (Kinin-9, Kallidin)

Bradykinin is the final product of the kinin system and is split from a serum α-2-globulin precursor by the kallikreins and also by trypsin or plasmin. Bradykinin reduces blood pressure by dilating blood vessels. In bronchial smooth muscles and also in the intestines and the uterus, bradykinin leads to muscle contraction. Bradykinin is also one the most potent known pain-inducing substances. BK causes hypotension, contracts smooth muscles and increases vascular permeability. It also plays a role in pain pathways and inflammation. BK antagonists are used for treating inflammations, pain, rheumatic arthritis, osteoarthritis, pancreatitis, rhnitis, asthma and gout. Bradykinin has a powerful influence in stimulating smooth muscle contraction, inducing hypotension, increasing blood flow and permeability of capillaries.

2.5 Vasopressin (VP)

This protein is also called antidiuretic hormone (ADH), adiuretin, vasotocin, pituitrin P and pitressin. It is a cyclic nona-peptide, synthesized in the hypothalamus and stored in the posterior lobe of the pituitary from which it is released into the circulation as necessary. Functions of VP include stimulation of ATCH release, improvement of the memory and learning capacity, reduction of the pressure in the pulmonary arteries and reduction of renin and ACE activity. Vasopressin regulates osmotic pressure in body fluids via a specific vasopressor receptor (V1). It has direct antidiuretic activity in the kidney, mediated by the antidiuretic receptor V2, and promotes re-adsorption of water in the distal convoluted tubules of the kidney. It also causes vasoconstriction in peripheral small blood vessels by stimulating smooth muscle cells in the cell walls to contract.

2.6 Angiotensins (I, II, III)

Angiotensin is a decapeptide originally found to be produced by kidney derived renin from an α-2 hepatic globulin. It is mainly known for its potent pharmacological activities. Angiotensin elevates blood pressure through its direct vasoconstrictor, sympathomimetic, and (through release of aldosterone) sodium-retaining activities.

Angiotensins are formed in biological fluids by the enzymatic cleavage of proteins. The species-specific enzyme renin, which can be generated by kallikrein from inactive prorenin, is responsible for the formation of angiotensin I (AT I) from globulin angiotensinogen (ATG). AT I that has no effect on the blood pressure, and is split by the membrane bound angiotensin-converting enzyme (ACE) to form angiotensin II (AT II).
Angiotensin II is a very potent vasoconstrictor substance and acts directly on the adrenal gland to stimulate the release of aldosterone. The inhibition of ACE results in a double hypotensive effect because both the formation of blood pressure raising AT II as well as the degradation of the blood pressure lowering kinin is inhibited. AT II agonists are used for treatment of shock and collapse in which a normal blood pressure could be restored as quickly as possible, while ACE inhibitors and AT II antagonists are applied as antihypertensive agents for treatment of hypertension.

2.7. Enkephalins (*Leu* and *Met*-enkephalin)

These compounds comprise the basis for the body's own pain fighting mechanisms. The enkephalins are found in many areas of the body. Changes in these compounds and their metabolism have been associated with different headache disorders.

The two 5-peptide enkephalins have been identified. One terminates in a leucine, and is known as *Leu*-enkephalin; the other terminates in a methionine, and is called *Met*-enkephalin. The enkephalins are relatively weak analgesics, which activate all opioid receptors, but appear to have the highest affinity for the d receptor. Apart from nervous tissue, enkephalins have been identified in many other organ systems, including the gut, sympathetic nervous system, and adrenal glands. In the CNS, enkephalins have been found in many areas but predominantly those associated with nociception (e.g. PAG and dorsal horn). Their pre-cursor molecule is proenkephalin and they are rapidly degraded by enkephalinase. Wondering why the human brain should have receptor sites for alkaloids from the opium poppy led to the discovery of a family of natural painkillers, the endorphins (from *endogenous morphines*). These substances are oligopeptides, containing from 5 to 30 amino acids.

2.8. Somatostatin (SS, SRIF)

Somatostatin, known also as somatotropin release inhibiting hormone (SIH) , is a peptide of 14 amino acids found in the hypothalamus and central and peripheral nervous system. Angiopeptin is a stable analog of somatostatin. Somatostatin (SRIF) is formed as prepro-SRIF. The main product of gene expression is pro-SRIF- (1-64), which is processed at the C-terminus to form SRIF-28 and SRIF-14. SRIF and SRIF-like substances have been found in the hypothalamus, central and peripheral nervous system, as well as the gastrointestinal tract. The main biological effect of SRIF is to inhibit the release of growth hormone, TSH, prolactin, CRH, insulin, glucagon, VIP, secretin, pancreatic polypeptide, gastrin releasing peptide, gastrin, CCK and motilin. A possible role for somatostatin in affective disorders is suggested by its low concentration in cerebrospinal fluid of patients with depression. Somatostatin in the brain might be involved in therapeutic effects of some of antidepressant drugs.

2.9. Bombesin (BN)

Bombesin, originally isolated from the skins of the amphibians *Bombina bombina* and *Boombina variegata variegata*, is a potent stimulant of gastric acid secretion and is shown to be strong biologically active in the central nervous system. These include regulation of the contraction of smooth muscle cells, induction of the secretion of neuropeptides and hormones. Bombesin increases the plasma levels of gastrin, CCK, glucagon, insulin, pancreatic peptide, VIP and many other gastrointestinal peptides. The C-terminal nonapeptide of bombesin has the minimum length with the maximum effect. Bombesin is used as a diagnostic aid in the gastrin stimulation test.

2.10. Neurotenisn

Neurotenisn is a 13 amino acid peptide isolated from bovine hypothalamus. It causes hypotension in the rat and its smooth muscle actions include relaxation of the rat duodenum and contraction of guinea pig ileum and rat uterus. Neurotensin may also act as a CNS neurotransmitter. Neurotensin is involved with memory function, as shown in the brains of Alzheimer's disease patients where there are deficits of this peptide in certain regions involved with memory function. This peptide may also be involved in the pathophysiology of Parkinson's disease and schizophrenia.

2.11. Oxytocin (ADH)

The posterior pituitary has two hormones, ADH (antidiuretic hormone, vaspopressin) and oxytocin which are medically important. Both of these hormones are small peptides containing nine (9) amino acids each. They are synthesized in the hypothalamus (supraoptic nucleous for ADH and paraventricular nucleus for oxytocin). Oxytocin stimulates contraction of uterine smooth muscle. It is secreted during labor to effect delivery of the fetus. Oxytocin also stimulates contraction of smooth muscle in the mammary glands (myoepithelial cells). Oxytocin causes smooth muscle contraction in the alveoli (small chambers) and larger sinuses of the mammary glands to make readily available milk, whose production has been induced by prolactin and estrogen, to the suckling infant. Oxytocin causes milk ejection, which is necessary for adequate lactation, but not milk production. Prolactin controls milk production in conjunction with estrogen.

2.12. Thyrotropin – Thyroliberin TSH (thyroid stimulating hormone)

A glycoprotein hormone consisting of two protein chains, one of which is identical with a subunit of Luteinizing hormone. Thyrotropin is produced in the anterior pituitary in response to thyrotropin releasing hormone (thyroliberin; thyrotropic hormone releasing factor or TRF). Thyrotropin stimulates the thyroid gland to secrete thyroid hormones such as thyroxin and triiodothyronin. These two hormones inhibit the secretion of TRF and thyrotropin. Thyrotropin stimulates secretion of prolactin and acts as a neurotransmitter in the central nervous system. Apart from its well-known physiological role, thyrotropin appears to be involved in the modulation of immune responses within the neuroimmune network. Thyrotropin enhances proliferation of lymphocytes stimulated by suboptimal concentrations of IL2 and enhances IL2-induced NK-cell activity. TSH also enhances production of superoxide anions by stimulated macrophages.

2.13. HRF (histamine-releasing factors)

This is a general term used for factors that induce the release of histamines from basophils and mast cells when stimulated with antigens or mitogens.

2.13.1 HRIF (Histamine release inhibitory factor)
This poorly characterized factor is a specific antagonist of histamine-releasing factors. It is produced by peripheral blood mononuclear cells (B-cells, T-cells, monocytes) upon stimulation with histamine or mitogens such as Con A. It inhibits HRF-induced histamine release from basophils and mast cells. One particular factor with HRIF activity is IL8 .

2.13.2 CTAP-3 Beta-Thromboglobulin (Beta-TG)
CTAP-3 (connective tissue activating protein-3) or beta-thromboglobulin is a protein of 8.85 kDa. Beta-thromboglobulin is stored in the Alpha-granules of platelets and released in large amounts after platelet activation. Beta-thromboglobulin is a strong chemoattractant for fibroblasts and is weakly chemotactic for neutrophils. It stimulates mitogenesis, extracellular matrix synthesis, glucose metabolism, and plasminogen activator synthesis in human fibroblast cultures.

Beta-thromboglobulin, its precursor, and its cleavage products influence the functional activities of neutrophilic granulocytes. Beta-thromboglobulin affects the maturation of human megakaryocytes and thus could play a role in the physiological regulation of platelet production by megakaryocytes.

2.14. and 2.15. Neurokinin A (NKA) or (Substance K) and Neurokinin B (NKB) or (Neuromedin K)

Neurokinins are found centrally in the spinal cord and in the sensorial nuclei of the brain stem and peripherally in the ends of the sensorial fibers. Neurokinins include Substance P (SP), neurokinin A (NKA), and neurokinin B (NKB). Similar compounds, which occur in cold-blooded animals, are called tachykinins. So far, three different receptors have been found for the neurokinins: NK1 for SP, NK2 for NKA, NK3 for NKB. Neurokinins play various roles in the regulation of cardiovascular system, pain pathway and inflammatory reaction. Neurokinin A and B belong to the tachykinin family. They are a more potent bronchio-constrictor than substance P and may regulate neutrophil recruitment in the lower respiratory tract. They arise from larger precursor molecules and exhibit functions such as vasodilatation, hypotension, extravascular smooth muscle contraction, salivation and increase of capillary permiability

2.16. Neuropeptide Y (NPY)

Neuropeptide Y (NPY) is present in the brain and in the peripheral nervous system along with other neurotransmitters. It has structural homology with pancreatic polypeptide (PP) and peptide YY (PYY). Its functions may include neurotransmission, neuromodulation, vasoconstriction, regulation of blood pressure, and appetite.

3. Assessment of Various Bioregulator Materials as Weapons of Mass Destruction

It was very hard to find in available literature all the data for all bioregulators especially for criterion: Agents known to have been developed, produced, stockpiled or used as weapons (in the tables - Weaponized). From a public health standpoint, bioregulators which are less known, must be evaluated and prioritized in order to assure appropriate allocation of the limited funding and resources that are often found within public health systems.

Our opinion is that if some bioregulator satisfies the bulk of the criteria, it should be recommended for inclusion in the list. As the list of bioregulators will be hard to define generally and for purposes of the future negotiations of various treaties, such as the Biological Toxin Weapon Convention, this paper proposes two tables of enlisted bioregulators with important criteria on the basis of which a decision can be made to include in or exclude from a list of the molecular agents (bioregulators). The following is a listing of the selection criteria for inclusion in these tables.

3.1 CRITERIA FOR SELECTION OF BIOREGULATORS AS TERRORISM AGENTS

1. High level of morbidity: higher rating (++) if clinical disease requires hospitalization for treatment including supportive care and lower rating (+) if outpatient treatment is possible for most cases.

2. High level of mortality or incapacity: agents with an expected mortality of $\geq 50\%$ were rated higher (+++), and with lower expected mortalities (21-49%=++, and <21%=+).

3. Stability in the environment after release (+).

4. Ease of production and transportation (+).

5. Likely methods for terrorism usage and high level of dissemination or contamination by aerosol for respiratory exposure (+++); contamination in quantities that could affect large populations (++); and dissemination potential for sabotage of food and water supply (+).

6. High toxicity or potency or low toxic dose: LD_{50} <0,000025 mg/kg (+++), LD_{50} from 0,000025 do 0,0025 mg/kg (++) and LD_{50} >0,0025 mg/kg (+).

7. High level of intoxication by variety route: per oral route (+), respiratory route (++), or both (+++).

8. Stockpiling of prophylactics and antidotal therapy (+).

9. Enhanced surveillance and education (+).

10. Difficult to diagnose or identify at the early stage or improved laboratory diagnostics (+).

11. Public perception: Public fear associated with an agent and the potential mass civil disruptions that may be associated with even a few cases of disease were also considered (+ to +++).

3.2 CRITERIA FOR SELECTION OF BIOREGULATORS AS WARFARE AGENTS

Agents known to have been developed, produced, stockpiled or used as weapons (+).
1. High level of dissemination potential for contamination a large area by aerosol for inhalathory exposure and in military significant quantities that could affect large populations (+++), and dissemination potential for contamination of food and water supply (++).

2. High toxicity or potency or low toxic dose: LD_{50} <0,000025 mg/kg (+++), LD_{50} from 0,000025 do 0,0025 mg/kg (++) and LD_{50} >0,0025 mg/kg (+).

3. High level of morbidity: higher rating (++) if clinical disease requires hospitalization for treatment including supportive care and lower rating (+) if outpatient treatment is possible for most cases.

4. High level of intoxication by variety route: per oral route (+), respiratory route (++), or both (+++).

5. High level of mortality or incapacity: agents with an expected mortality of ≥50% were rated higher (+++), and with lower expected mortalities (21-49%=++, and <21%=+).

6. No effective prophylaxis and therapy commonly available and widely in use (+).

7. Stability in the environment (+).

8. Difficulty to diagnose/detect or identify at the early stage (+).

9. Ease of production and transportation (+).

After reviewing these criteria and applying them to the bioregulators, the following two tables were derived to determine potential of bioregulators as either a conventional biological warfare (Table 1) or as a potential terrorist weapon (Table 2).

Table 1. Assessment of bioregulators as conventional biological warfare agents.

Bioregulator	(1) Weapo-nized (+)	(2) High level of dissemi-nation (+++)	(3) High tox-icity (+++)	(4) High morbi-dity (++)	(5) Intoxication by variety of route; per oral route, respiratory route, or both (+++)	(6) High level of incapacity/mortality (+++)	(7) No effective prophylaxis and therapy (+)	(8) Stability in the environ-ment (+)	(9) Difficulty of detection/identifi-cation (+)	(10) Ease of produc-tion (+)	Total (19)
1. Endorphins (α, β, and δ-Endorphin)	+	+++	+++	++	+++	+++	+	+	+	+	19
2. Substance P (SP) (Neurokinin)	+	+++	+++	++	+++	+++	+	+	+	+	19
3. Endothelins (ET-1, ET-2, ET-3) or Sarafotoxins (S6a, S6b)	+	+++	+++	++	+++	+++	+	+	+	+	19
4. Bradykinin (Kinin-9, Kallidin)	+	+++	+++	++	+++	+++	+	+	+	+	19
5. Vasopressin (VP)	+	+++	+++	++	+++	++	+	+	+	+	18
6. Angiotensins (I, II, III)	+	+++	+++	++	+++	++	+	+	+	+	18
7. Enkephalins (Leu- and Met-enkephalin)	+	+++	+++	++	+++	++	+	+	+	+	18
8. Somatostatin (SS, SRIF)	+	+++	++	++	+++	++	+	+	+	+	17
9. Bombesin (BN)	+	+++	++	++	+++	++	+	+	+	+	17
10. Neurotensin	+	+++	++	++	+++	++	+	+	+	+	17
11. Oxytocin	+	+++	++	++	+++	++	+	+	+	+	17
12. Thyroliberins (Thyrotropin)	+	+++	++	++	+++	++	+	+	+	+	17
13. Histamine releasing factors (HRF): - Histamine release inhibiting factor (HRIF) - CTAP-3 Beta-Thromboglobulin (Beta-TG) - Neutrophil-activacting factor (NAF) - Stem cell factor (SCF)	+	+++				++	+	+	+		17
14. Neurokinin A (NKA)/Substance K (SK)	-	+++	++	++	+++	++	+	-	+	+	15
15. Neurokinin B (NKB)/Neuromedin K	-	+++	++	++	+++	++	+	-	+	+	15
16. Neuropeptide Y (NPY)	-	+++	++	++	++	++	+	-	+	+	15

Table 2. Assessment of bioregulators as terrorism agents.

Bioregulator	(1) High morbidity	(2) High level of mortality/ incapacity	(3) Stability in the environment	(4) Ease of production	(5) High level of dissemination	(6) High toxicity	(7) High level of intoxication	(8) No effective prophylaxis and antidotal therapy	(9) Enhanced surveillance and education	(10) Difficulty of detection/ identification	(11) Public perception	Total
	(++)	(+++)	(+)	(+)	(+++)	(+++)	(+++)	(+)	(+)	(+)	(+++)	(22)
1. Endorphins (α, β, and δ-Endorphin)	++	+++	+	+	+++	+++	+++	+	+	+	+++	22
2. Substance P (SP) (Neurokinin)	++	+++	+	+	+++	+++	+++	+	+	+	+++	22
3. Endothelins (ET-1, ET-2, ET-3) or Sarafotoxins (S6a, S6b)	++	+++	+	+	+++	+++	+++	+	+	+	+++	22
4. Bradykinin (Kinin-9, Kallidin)	++	+++	+	+	+++	+++	+++	+	+	+	+++	22
5. Vasopressin (VP)	++	+++	+	+	+++	+++	+++	+	+	+	+++	22
6. Angiotensins (I, II, III)	++	+++	+	+	+++	+++	+++	+	+	+	+++	22
7. Enkephalins (*Leu*- and *Met*-enkephalin)	++	+++	+	+	+++	+++	+++	+	+	+	+++	21
8. Somatostatin (SS, SRIF)	++	+++	+	+	+++	+++	+++	+	+	+	+++	21
9. Bombesin (BN)	++	+++	+	+	+++	+++	+++	+	+	+	+++	21
10. Neurotensin	++	+++	+	+	+++	+++	+++	-	+	+	+++	21
11. Oxytocin	++	+++	+	+	+++	+++	+++	+	+	+	+++	21
12. Thyroliberins (Thyrotropin)	++	+++	+	+	+++	+++	+++	+	+	+	+++	21
13. Histamine releasing factors (HRF):												
- Histamine release inhibiting factor (HRIF)												
- CTAP-3 Beta-Thromboglobulin (Beta-TG)												
- Neutrophil-activating factor (NAF)												
- Stem cell factor (SCF)												
14. Neurokinin A (NKA)/Substance K (SK)	++	++	-	+	+++	+++	+++	+	+	+	+++	20
15. Neurokinin B (NKB)/Neuromedin K	++	++	-	+	+++	+++	+++	+	+	+	+++	20
16. Neuropeptide Y (NPY)	++	++	-	+	+++	+++	+++	+	+	+	+++	20

Potential terrorism and warfare bioregulators were given with an expected mortality of ≥50% were rated higher (+++) than agents with lower expected mortalities (21-49% = ++, and <21% = +).

Bioregulators with higher rating (++) for morbidity are if clinical disease required hospitalization for treatment (including supportive care), and with lower rating (+) if outpatient treatment was possible for most cases.

Bioregulators received + to +++ for dissemination potential based on their likely methods of contamination a large area by aerosol for respiratory exposure (+++), on contamination in quantities that could affect large populations (++), and sabotage on food and water supply (+).

High level of intoxication by variety route is showed according of the kind of exposure: per oral route (+), respiratory route (++), or both (+++).

Bioregulators also were ranked based on any special public health preparedness that was required including: stockpiling of therapeutics (+), enhanced surveillance and education (+), and improved laboratory diagnostics (+).

Public fear associated with an agent and the potential mass civil disruptions that may be associated with even a few cases of disease were also considered (+ to +++).

4. Conclusions

The results of this effort, as shown in the preceding two tables, shows that it is very hard to make a final decision on criteria and the final list of the molecular agents (bioregulators) for their potentials as weapoins of mass destruction. Although many biological agents such as bioregulators can be used to cause illness, there are only a few that can truly threaten civilian populations on a large scale. If released upon a civilian population, these agents would pose the most significant challenge for public health and medical responses.

The above criteria for ranking potential bioregulators and listing of them of greatest public health concern could be used for determination of priority biological threat agents for national public health preparedness efforts for bioterrorism. Having a defined method for evaluating biological threat agents allows for a more objective evaluation of newly emerging potential threat agents, as well as continued re-evaluation of established threat agents. Using this prioritization method can help focus public health activities related to bioterrorism detection and response and assist with the allocation of limited public health resources.

40

5. References

1. Alibek, K.; Handelman, K., Biohazard: The chilling True Story of the Largest Covert Biological Weapons Program in the World-Trade from the Inside by the Man Who Ran It, Random House: New York, 1999.

2. Rotz, D. L., Khan, S. A., Lillibridge, R. S., Ostroff, M. S. and Hughes, M. J., Priotritizing potential biological terrorism agents for public health preparedness in the united states: Overview of evaluation process and identified agents, National Center for Infectious Diseases, Centers for Disease Control and Prevention, Proceedings of CBMTS III, 2000.

3. A FOA Briefing Book on Chemical Weapons, Theath, Effects and Protection. SIPRI, Oxford University Press, Oxford, (1987).

4. Grant J.A. et al, Histamine-releasing factors and inhibitory factors. International Arch. Allergy Appl. Immunology, 1991, 94: 141-3.

5. Kuna P. et al, Further characterization of histamine releasing chemokines present in fractionated supernatants derived from human mononuclear cells. Clin. Experimental Allergy, 1996, 26(8): 926-33.

6. Chen L.I. et al, The interaction of insulin and angiotensin II on the regulation of human neuroblastoma cell growth. Mol. Chem. Neuropathology, 1993, 18: 189-96.

7. Clyne C.D. et al, Angiotensin II stimulates growth and steroidogenesis in zona fasciculata/reticularis cells from bovine adrenal cortex via the AT1 receptor subtype. Endocrinology, 1993, 132: 2206-12.

8. Wolf G. and Neilson E.G., Angiotensin II as a hypertrophogenic cytokine for proximal tubular cells. Kidney International Suppl., 1993, 39: S100-7.

9. Yamamoto Y. et al, Angiotensin III is a new chemoattractant for polymorphonuclear leukocytes. Biochemical and Biophysical Research Communications, 1993, 193: 1038-43.

10. Rozengurt E., Neuropeptides as cellular growth factors: Role of multiple signaling pathways. European Journal of Clinical Investigation, 1991, 21: 123-34.

11. Woll P.J. and Rozengurt E., Two classes of antagonist interact with receptors for the mitogenic neuropeptides bombesin, bradykinin, and vasopressin. Growth Factors, 1988, 1: 75-83.

12. Burch R.M. and Kyle D.J., Recent developments in the understanding of bradykinin receptors. Life Sci., 1992, 50: 829-38.

13. Hess J.F. et al, Cloning and pharmacological characterization of a human bradykinin (BK-2) receptor. Biochemical and Biophysical Research Communications, 1992, 184: 260-8.

14. Leppaluoto J. and Ruskoaho H., Endothelin peptides: biological activities, cellular signaling and clinical significance. Ann. Med., 1992, 24: 153-61.

15. Miller R.C. et al, Endothelins - from receptors to medicine Trends in Pharmacological Sciences, 1993, 14: 54-60.

16. Moore, G. J., Designinig peptide mimetics. Trends in Pharmacological Sciences 15: 124-129 (1994).

THE PROBLEMS OF CHEMICAL TERRORISM

CHRISTOPHER D. DISHOVSKY
Military Medical Academy, St. G. Sofiisky, 3, Str., Sofia , Bulgaria

Abstract

The use of sarin in the terrorist acts in Matsumoto city and the Tokyo underground attracted the attention of the scientific society to the chemical weapons (CW). But release of chemical materials from industrial sites poses also hazard and risks to military forces, emergency response person and citizens alike. There will be problems in detection and protection of chemical agents. Military forces are trained and equipped to survive and operate in toxic chemical environment. It is important to determine the capabilities of military and civil emergency personnel and units to respond to chemical terrorism in terms of existing equipment, skills, organization and procedures. The military medical corps of Balkan countries can discuss the possibility to help each other in the case of chemical terrorist acts. It is possible to be created Balkan Pharmaceutical Stockpile program including chemical antidotes and life support medications and equipment.

1. Introduction

The recent use of sarin in the terrorist acts in Matsumoto city and Tokio underground was considered by a number of specialists like a new era in terrorism. It removed any doubts about the possibility of using chemical weapons by terrorists. Today it is clear that terrorism can be not only a state policy, but can also be realized by separate individuals, groups or organizations.

In addition to chemical weapons, terrorists can use different toxic chemicals from chemical industry, agriculture or products released from terrorist acts on industrial facilities. Some specialists include different toxins in this group of terrorist agents.

2. Discussion

Chemical weapons have a number of advantages that make them a priority among terrorist agents. They are:

- relatively easy and cheap to produce;

- easy to access precursors of chemical weapons, which are routinely used in industry and in everyday life;
- precursors and toxic chemicals are easy to carry and transport because of difficulties to monitor and control their movement;

- resulting damage after an attack is considerable in amount and content;

- their fast effect requires an emergency response which makes it difficult to execute the rescue operation effectively;

- their psychological impact will extend far beyond the actual size of damage or number of casualties in time;

P. Stopa and Z. Orahovec (eds.), Technology for Combating WMD Terrorism, 41-44.
© 2004 *Kluwer Academic Publishers. Printed in the Netherlands.*

- the arsenal of chemical materials that can be used by terrorists is practically unlimited.

Chemical agents that might be used by a terrorist include:

-chemical weapons that are already known;

-chemical agents, which may be used as chemical weapons with the help of a new technology;

-unknown chemical agent, which may be used as chemical weapons;

-chemical agent, which cause public panic and social disruption;

-chemicals commonly used in the industry;

- pesticides;

-toxins;

-chemicals, which may be obtained after explosion, fire and other incidents in industry or transportation of chemicals.

The response system in case of chemical terrorist acts is a problem of all national institutions. With regard to this the civilian institutions have some advantages (for example in Bulgaria):

-well organized civil defense;

-identification and classification of dangerous substances and chemical facilities;

-documentation of industrial processes and products;

-development of risk models;

-environmental monitoring;

-well organized health system (sufficient number of hospitals, physicians and pharmacists);

-Development of environmental protection in line with the European Union (EU) criteria;

-developed communication system;

-experience in studying and producing antidotes and individual protective equipment.

On the other hand, some countries have civilian institutions that have disadvantages:
-lack of stockpiling of antidotes and other life-saving aids;

-less teaching and training activity/programs for fight against chemical terrorism at all levels and

structures of society (including physicians);

-lack of identification and designation of specific hospitals and alternative health care facilities for managing of large number of patients after chemical terrorist acts.
The issue of chemical terrorism has direct relevance for the army. On one side, a military unit can be the target of a terrorist act. On the other, as practical experience shows, army units and the army medical corps are involved and take active part in coping with industrial accidents and natural disasters and the management of their consequences.

The army, in general, is perhaps in a better position than civilians to act in chemical counter-terrorist operations for the following reasons: the higher level of training,; availability of chemical defense equipment; antidotes; and devices for detection and decontamination. The problem is, however, that all this potential and readiness is prepared for effective use in conditions of chemical warfare and not for civilian response.

The question arises whether the army units involved have developed action plans in case of chemical terrorist acts and attacks either on them or in the vicinity of their location. Such plans for the army should be an integral element in the government counter-terrorist policies and politics of each country.

According to the conclusions reached by a panel of international experts, many countries still lack an overall concept and vision of the army's participation in fighting and eliminating the outcome of chemical terrorism within that country itself.

Within the overall counter-terrorist coordination among different national agencies and units, the planning and preparation of the army units and their medical services should be focused on the following specific points:

- risk assessment for the use of chemical agents as terrorist agents with particular attention to toxic industrial chemicals and toxins.

- update the assessment of the effective toxic levels that should cover both the known chemical weapons in view of the modern technologies of their use and toxic compounds and chemicals of industrial origin.

- assessment of the available capability and contemporary technological devices for the detection and identification of a broad range of chemical compounds.

-modernization and optimization of individual protection with particular focus on respiratory protection and protective clothing.

-inventory and assessment of the available means for medical treatment of chemical intoxications. Assessment of the required amounts and types of antidotes (in view of the broader range

-assessment of the available means for indication and control of chemical contamination of potentially toxic agents) and their update with development and introduction of new compounds. and the effectiveness of decontamination. This should consider the broader range of potentially toxic agents and the available state-of-the-art technologies.

The training of personnel acquires particular significance in the preparation of the army and its medical corps to counteract chemical terrorism. It should incorporate and implement the latest achievements of computer simulation and virtual reality technologies.

The experience of other countries (e.g.: Czech Republic) shows that the military medical teaching and research communities can play a considerable role in raising public awareness and preparing the population and civic organizations for the fight against chemical terrorism.

The effective preparation of the army for action in chemical and other terrorist attacks is an expensive and long-lasting continuous process that can be improved and made more productive with the joint coordinated efforts of neighbor countries. There are a number of opportunities and unused potential in that respect:

-coordination of resource utilization and trans border mutual aid in terrorist acts near the borders of neighboring countries.

-use of available special military medical units in emergency situations;

-common stockpiling of antidotes and other life-saving aids.

-unified notification and information system for the applied primary medical treatment.

-effective triage and transport to the nearest specialized medical facility.

-joint exercises of the medical corps of the fellow countries in managing and eliminating the consequences of chemical terrorism.

3. Conclusion

The basic aspects of a common national plan for the fight against chemical terrorism are:

-participation of all institutions;
-regional/international cooperation.

The main topics of this plan should be:

-protection of human life from damage caused from chemical agents after chemical terrorist acts;
-defense against losses on the environment;
-preventing material losses.

SUPPORTING SCIENCE AND TECHNOLOGY:

**COMMENTARY
AND
TECHNICAL
APPROACHES**

PREFACE

Once the threat from weapons of mass destruction has been identified, one needs to develop a capability to counteract these weapons. In this section, we investigate approaches from a both a policy and technology approach.

One of the ways to reduce the threat from these materials is to limit the information that is available to rogue states, terrorist organizations, or individual terrorists. There is a tradeoff, however, since we must maintain open and responsible communication within the scientific community.

Secondly, we must assemble technologies to be able to detect, protect, and mitigate the effects of a WMD attack. The technologies available for biological detection and identification area broad, but we still must come to a consensus as to their implementation in the field and in the reference laboratory. A review if currently available technologies for biological detection are presented along with some potential strategies for their implementation in the field. Some practical experience with candidate generic detectors is discussed. A new strategy for the development of probes for identification purposes is also shown.

Chemical agent detection is more mature, but one must realize that we need several types of detectors for agent in their various states. There are several inexpensive and effective technologies that one may use for chemical agent detection.

In addition to technologies, scientific support is also needed to help counteract the effects of these agents. An excellent example of this type of work is contained within this area.

After an attack, one needs to protect and mitigate the effects. This extends to the military, civilian responders, and the civilian population in general. The experiences of the Czech Republic in the area of protection and personal decontamination are presented as a paradigm for this area.

Lastly, we must look to the future and what technologies have to offer. Future technologies and their impact on the future response are presented. The crystal ball shows that arrayed sensors, telemedicine, and improvements in the human interface will have a positive impact on the response to a WMD event in the future.

CHALLENGES TO THE DESIGN OF NEW DETECTION DEVICES

MARIA JOSE ESPONA
Presidency of the Republic, Argentina
Sanchez De Bustamante 2173
Piso 17 Depto J C.P. 1425
Capital Federal, Argentina

Abstract

As in a DNA double helix, the progress in the capacity to design detection devices runs parallel with the advances in the creation of genetically modified microorganisms. The links of these two spirals is constituted by extensive and freely available information on new detector development and research and experimentation with microorganisms. In the near future, this phenomenon could lead to a race between detection device designers and proliferators or terrorist groups that may gain access to the information and technology required to counter them. This paper highlights the relevance of protecting both critical know-how and intangible technology in order to deter or limit the capacity of proliferators in the biological arena.

1. Introduction

This paper is focused on the current and future situation in the design of new detection devices and to what extent freely available information may affect their development.

This study was performed by analyzing open sources of information.

1.1 WHAT DOES DETECTION MEAN?

Before dealing with the specific issues of this paper, it is convenient to define what the word "detect" means in this context: to identify the presence of uncommon pathogenic microorganisms, whether by quantity or space/time location, that may be harmful to health conditions in human beings, animals or crops. In order to identify an anomaly in the system, that in this case would mean identifying the presence of a biological agent, we may follow basically two paths: to know thoroughly the characteristics of the environment (where the detection will be performed) or the biological warfare agents (BW agent). Considering these two aspects, the detection of BW agents represents a great challenge, since it is almost impossible to know and characterize every possible environment, as well as to distinguish every potential BW agent (it is even a greater challenge when considering that they can be genetically modified).

1.2 PROGRESS IN SCIENCE AND TECHNOLOGY VS. DEVELOPMENT OF ENHANCED BW ARSENALS

In the past several years, biological sciences in general -such as genetic engineering and biotechnology- and the study of microorganisms in particular -microbiology- underwent a significant growth. What do these scientific disciplines, the design of detection devices and the development of biological weapons have in common? Information and know-how. We have to always bear in mind that the information we are referring to is mostly public and freely available in scientific and industrial sectors, principally in pharmaceuticals.

The control of these two sectors is particularly complex: the academic sector is closely related to the advance of science and, through its application, it saves thousands of lives. Therefore, controlling the academic sector could result in worse health care conditions and a lower population growth rate (availability of food, access to low-cost health care centers, access to drinking water resources, treatment of illnesses, etc.).

P. Stopa and Z. Orahovec (eds.), Technology for Combating WMD Terrorism, 49-52.
© *2004 Kluwer Academic Publishers. Printed in the Netherlands.*

As regards to the industrial sector, it must be noted that there is a significant degree of self-control in order to protect the industrial secret and the investment made to develop a product. Yet, very often the methodology used to develop a product or the principles on which it was developed are revealed when offering the product. The protection of information is particularly important when dealing with the development of detection systems : if the principles on which these systems are based on to operate are revealed, the information about how to harm them is implicitly revealed as well.

On the other hand, historically when referring to BW agents they included natural microorganisms and maybe a specially selected pathogen strain, as an example of advanced development. Nevertheless, the current progress of genetic engineering and biotechnology extend the possibility of producing customized BW agents. If we were to divide potential biological agents into groups, they would be:

1. Known natural microorganisms;

2. Genetically modified microorganisms in order to:
      a. avoid their identification;
      b. provide them with characteristics they lacked (for instance, a flu virus with the genes of a hemorrhagic fever, causing a new type of illness, such as an "hemorrhagic flu"),

3. New agents (for instance, those that have not been discovered yet because they remain in a virgin rainforest without having come into contact with human populations).

In cases 1 and 2b, the detectors can be programmed in order to identify the BW agents. But for 2a it is necessary to look for another way of determining if there is any microorganism that has the potential to cause damage. The situation in case 3 is similar to case 2a, with the exception that if case 2a involves a microorganism modified not to be detected, once identified, it is possible to develop a vaccine or treatment. Yet, in case 3, many trials might be necessary until the correct countermeasure is found, which require a lot of valuable time.

As regards what kind of know-how to protect, the information on applied methods is very often more sensitive than the data that refers to specific cases. For instance, the method to make genetically modified microorganisms is not new: it is at least 10 years old and it is virtually impossible to revise all the methods revealed in magazines, books, scientific exchanges, joint projects, symposia and conferences.

2. Detection Devices

By and large, working to enhance detection and protection systems comes as a response to the perception of growth and/or upgrade of an enemy arsenal. These activities are similar in both sides of the equation, thus generating both a defensive and offensive armament race.

In the case of biological weapons, the situation is far more complex: in a first stage, biological weapons and a means of protection were developed, leaving detection aside. Later and with the increasing biological threat posed by non-state actors, relevance was given to the production of detection devices capable of providing early warning and identifying in real time the presence of BW agents. In fact, it may be asserted that it is only now that we are very close to achieving a mobile real-time detection system.

In the near future, the problem that the developers of detection devices must face will be *the great availability of information for the development of genetically modified biological agents, in particular those altered to avoid their detection.*

3. Must Critical information Be Protected?

The current challenge in the field of non-proliferation is to control the flow of information without affecting the progress of science and medicine and at the same time avoiding that this critical information might be acquired by groups or countries with the intention of developing biological weapons - even if it is not in the near future. The protection of information is not new to the scientific community. A recent widely-known reference is what occurred during the Cold War when the Government of the United States tried to constrain information exchange in some areas of mathematics and the physical sciences that might have aided Soviet nuclear weapons development. But, even at the height of the Cold War, the National Academy of Sciences (NAS) concluded that a higher security level would be achieved by open pursuit of scientific knowledge than by curtailing free exchange of scientific information.

In this context, the American Society of Microbiology has adopted policies and specific procedures for its publications, so that the developments that are of national security concern, particularly those dealing with select agents, be modified in such manner to avoid the transmission of relevant know-how on the production of WMD. The negative side is that it affects the dynamic of the progress of science, since it is almost impossible to prove the results published by other scientists because not all the information is known.

But on the other hand, other institutions like the Massachusetts Institute of Technology (MIT) reject classified research because it conflicts with their educational missions.

Another issue is to whom not to give access to the information. First, the "bad boys" come up - countries comprising the Axis of Evil, countries suspected of developing WMD programs, individuals or organizations who may be linked to terrorist groups, et alia. Thus, the number of countries of proliferation concern may rise from 3 to dozens as well as the number of entities to control.

Thus, the long-standing tension between openness in science and the protection of national security, continues.

The wars against terrorism and infectious diseases are global. If governments move towards restraining the flow of information across national boundaries, there will be an inevitable clash with the academic research community that is increasingly seeking international collaborations and partnerships. Limiting information exchange could slow the discovery of vaccines and drugs to treat infectious diseases, including those needed to defend the population against bioterrorism.

In the case of information control, the result would not be valid if it is performed by one country only, even if that it were the United States, since life sciences are in an advanced level in many places around the world. It is important to bear in mind that it is a global problem in the globalization era. Currently, security is presented as the clash between good and evil: developments for the progress of science, medicine and defense vs. development of increasingly effective BW agents that may break the defenses.

4. Conclusions

Governments and professional organizations must proceed jointly in avoiding the widespread dissemination of potentially useful information for the development of BW agents. It is worthless that a state or an organization alone takes measures to avoid the diffusion of critical information. The control over know-how transfers must be done through a multilateral consensus approach, without curtailing any individual freedom nor originating any xenophobic behavior.

In this framework, the strengthening of the Biological Weapons Convention gains great importance as a forum where the criteria for the transfer of materials and information could be agreed.

Balancing openness of scientific communication with classification, sensitive homeland security information and national security will be an issue difficult to solve. In this context, the scientific community may take the lead to develop self-policing procedures that protect national security and permit the advancement of science needed for the protection of public health.

In the case of detection devices in particular, should the information on BW detection devices technology or the identification phases be published, the system may become vulnerable: if an enemy knew that soldiers from certain country carried a specific detection device, the enemy would use other agents (that the device cannot detect) or different means to infect the soldiers.

It is extremely important to acknowledge the value of the information to be transmitted and to be aware of its possible misuse.

5. Acknowledgment

To Peter Stopa for his invitation to participate in this Conference and to the NATO Scientific Affairs division for their financial support.

6. References

1. Ronald M. Atlas; National Security and the Biological Research Community; www.sciencemag.org/cgi/content/full/298/5594/753

2. Scientific Communication and National Security: A Report Prepared by the Panel on Scientific Communication and National Security, Committee on Science, Engineering, and Public Policy, National Academy of Sciences, National Academy of Engineering, and Institute of Medicine (National Academy Press, Washington, DC, 1982).

3. J. Cello, A. V. Paul, E. Wimmer, Science 297, 1016 (2002); published online 11 July 2002 (10.1126/science.1072266).

4. http://journals.asm.org/misc/Pathogens_and_Toxins.shtml

5. Leo, Technol. Rev. (June 20), 60 (2002).

6. A recent report recommended that MIT ban classified research on its main campus to protect its educational mission--faculty could conduct such research on MIT's Lincoln campus. In the Public Interest: Report of the Ad Hoc Faculty Committee on Access to and Disclosure of Scientific Information (Massachusetts Institute of Technology, Cambridge, MA, 12 June 2002).

7. E. B. Skolnikoff, Chron. High. Educ. (10 May, 2002).

8. Peg Brickley, New Antiterrorism Tenets Trouble Scientists, New Scientist, (Oct. 28, 2002)

9. Gregory A Petsko; An Asilomar moment; http://genomebiology.com/2002/3/10/comment/1014

TECHNICAL APPROACHES TO BIOLOGICAL AGENT DETECTION

PETER J. STOPA
JOHN WALTHER
US Army Edgewood Chemical Biological Center
5183 Blackhawk Road
AMSRD-ECB-ENP-MC, E3549
Aberdeen Proving Ground, MD 21010-5424 USA

JEFF MORGAN
Applied Ordnance Technology, Inc.
25 Center Street
Stafford, VA 22554

1. Introduction

One of the keys to deterring the use of biological weapons is real time detection and warning. In the event that these agents are utilized, it is even more important to classify and eventually identify the type of agent so that appropriate countermeasures can be initiated. Several systems have been developed to provide detection and alarm of a biological warfare (BW) agent attack. These first generation systems detect the characteristics of an aerosol in order to measure changes in the aerosol content against a background. This may be indicative of a man-made (not naturally-occurring) event that could indicate a possible attack with a biological agent. These first generation systems then use antibodies to provide a means of characterization of the aerosol for specific types of biological materials. This approach presupposes a knowledge of what an adversary may have in their arsenal. With the advent of genetic engineering and thus the potential to design agents that may defeat current detection and identification strategies, additional or alternative signatures need to be exploited to reliably indicate a possible biological attack. This paper explores some additional signatures that can be used. Some approaches to improving medical intervention are also discussed.

1.1 UNKOWN AGENTS

What is meant by unknown agents? They are agents that are outside the sensor's library of known agents. They could be different agents or ones that have been genetically modified. This is even of more significance now that so many genetic sequences have been defined and so much information is available on the Internet. It is conceivable that in the not so distance future one could pick specific sequences to develop biological agents with defined properties. Unknowns could also include encapsulated agents and/or next generation BW agents, ones that are not readily recognizable with today's reagents.

1.2 RATIONALE

The rationale for the importance of being able to detect and identify these agents is as follows. First of all, we need to differentiate between natural and non-natural (i.e., man-made events). A second reason is that we need to provide warning to troops and the civilian population that an attack may have occurred so that the potentially affected people can implement and maintain a protective posture. Likewise, one also needs to know when to retire from this posture. In terms of consequence management, one needs to be able to initiate the appropriate medical treatments for the affected population. And lastly, one needs to know the areas that need to be decontaminated and how effective that decontamination is.

2. Strategies for Detection

2.1. TRIGGER AND DETECTOR STRATEGIES

A strategy for a response to a biological attack is illustrated in Figure 1. It certainly would be beneficial to recognize a biological weapon when it is still in aerosol form. This would be important to "Detect to Warn" and

P. Stopa and Z. Orahovec (eds.), Technology for Combating WMD Terrorism, 53-66.
© 2004 *Kluwer Academic Publishers. Printed in the Netherlands.*

also to trigger the sensor to collect a sample for further analysis. Logically, the next step would be to classify it as a bacteria, virus, protein toxin, or non-protein toxin. As a final step, you would want to identify it, which is important from a "Detect to Treat" perspective. Identification may also be necessary to prevent the outbreak of a disease or control secondary infections, especially if it is potentially communicable, such as plague, smallpox, Marburg, or Ebola virus. Identifying the agent could also be important for determining the proper medical response. Rapid tests could be performed to determine susceptibilities to various treatment regimes.

Currently particle size is the most widely used parameter that triggers an alarm, although technologies that measure particle shape, fluorescence from biological markers (tryptophan or NAD/NADH), or ATP luminescence are being developed and implemented into detection systems.

There are additional signatures that can be exploited. For example, elemental analysis can be performed to evaluate if a change in the ratio of various elements has occurred. Organic signature analysis can be performed to determine if materials that are consistent with various propellants, encapsulants, and aerosol additives are present. Changes in either the inorganic (elemental) or organic signatures could be indicative that a biological agent or agents may have been used. These parameters can be used as an initial approach to determine if the background aerosol characteristics have changed; however, further characterization is warranted. A summary of these approaches is illustrated in Figure 2.

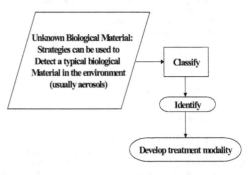

FIGURE 1. A strategy for response to a biological agent attack.

As can be seen by the figure, there are a variety of strategies that one could employ in trying to classify the components of an unknown biological agent. First, there are several parameters that may be exploited to determine if an agent is either natural or man-made. These are shown in Figure 3. Particle size is currently used by many fielded systems.   Likewise, particle shape, especially the frequency of occurrence of a particular shape, could serve as a trigger.

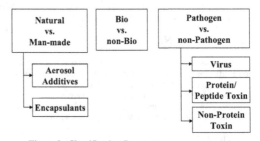

Figure 2. Classification Parameters

Elemental analysis is currently being considered by several groups as a means to determine whether a biological attack has occurred. For example, by determining the ratio of 2 or more elements, such as sodium, potassium, cobalt, zinc, etc., one might be able to determine if there has indeed been an attack. Calcium could be an indicator of spores, and nickel could be an indicator of botulinum toxin.   There could also be additives in the mix, such as silica, diatomaceous earth, or cellulose, which ordinarily are not found in the atmosphere. The propellants that are used in a biological dissemination device may also have measurable signatures.   An adversary may choose to encapsulate their weapon for a number of reasons. Encapsulants could be used to provide protection, to make the weapon more effective, or perhaps to cause false readings in sensors. Some technological approaches for the implementation of these strategies are shown Figures 3, 4, and 5.

2.2 CHANGES IN BIOLOGICAL FLUX

For example, it might be useful to see if there is a dramatic change in the biological flux of the environment. This is currently performed on existing platforms through the use of DNA measurements by flow cytometry and changes in the flux of Adenosine-Tri-Phosphate (ATP) by luminescence techniques. However, there are

Parameter	Rationale	Test	Status
Particles	Most man-made aerosols are > 3 µm	Particle-size Analysis	Available (Field)
	May see uniform distribution of similar particle shapes	Particle-shape Analysis	Available (Field)
Elemental Analysis	May see "unnatural" elements or ratios of elements.	Spectrometric analysis	Available (Field) – Photoionization detectors (PO₄, SO₄)\n\nIn Development – Sondebido Systems (Metals)
	May see spectra from silica, cellulose, or other organics	Active/Passive Standoff	Requires investigation
Propellants	May see propellants or other gases	Standoff; Gas Analyzers; IR FLIR; FTIR; Spectrometry	Requires investigation

Figure 3. Approaches for the determination of a suspect biological attack may contain a natural or man-made biological agent.

Parameter	Rationale	Test	Status
Silica Particles	Silica particles were (are) used to improve aerosol dissemination	Silica Analysis	Available (Field) – Various commercial manufacturers.
Cellulose Particles	Could be added to act as a stabilizer/filler.	Mass Spectrometry; Py-GC-IMS; Standoff	Under development
Protein Content	Protein content in air is low. Could be used as an additive or agent itself.	Standard Protein Determinations	Available (Field) – Various protein detection kits are routinely used.
	Could detect species markers from either additives or cells.	Histocompatibility Antigens	Available (Lab) – Used in FCM; could be adapted to other applications.
Residual Process Materials	Detect residual heme content from culture media, cells, etc.	Luminol Luminescence	Concept used during US BDWS program.
	Would have signatures from residual culture additives (e.g., vitamins and growth factors).	Mass Spectrometry	Available (Field)

Figure 4. Methods to detect aerosol additives or contaminants.

Parameter	Rationale	Test	Status
Particle Measurements	Encapsulated particles may have larger sizes, changes in refractive index, etc. that could be detected by scatter.	Refractive Index Shape Density Particle Size	Technology exists.
Absorbance	Encapsulants may impact the absorption spectra of biological materials.	Absorbance Spectra (UV/Vis/Ir)	Technology exists.
Composition	Some encapsulant materials may have unique signatures.	Mass Spectrometry Py-GC-IMS Elemental analysis	Technology exists.

Figure 5. Detection of encapsulant materials in biological agents.

Parameter	Rationale	Test	Status
DNA / Protein Determination	Unknown agents will probably be present in particles that have measurable DNA / protein.	Flow Cytometer	Available (Field)
	Same as above; however measure total flux rather than only particulates.	Fluorometer	Available (Field)
Heme Determination	Biological materials will have heme present	Luminol Luminescence	Available (Field) – US BDWS Program.
Viability Determination	Viable biological materials have ATP present.	ATP Luminescence	Available (Field) – System Demonstrated
	Measure presence of lipases.	Fluoroscein Diacetate	Available (Field) – Routine reagent used in FCM and fluorescence.
Image Analysis	Perform Gram Stain with automated morphological analysis.	Conventional Microscopy	Gram Stain is widely used.

Figure 6. Methods to determine whether particles in an aerosol are biological in origin.

additional parameters that can be measured. For example, simultaneous DNA and Protein measurements have been shown to be effective to measure changes in the biological flux. Other parameters, such as heme measurements, viability changes, and a version of the Gram stain may also prove to be useful. Examples of this approach are seen in Figure 6.

2.3. PATHOGENICITY/VIRULENCE DETERMINATION

If a change in the bioflux is determined to be significant, the next step would be to determine whether this change is a biological material that is dangerous, i.e., may cause death or poses a threat to health. Thus we need to make a determination as to whether or not the material is a pathogen or a non-pathogen.

Various approaches can be used to make such a virulence determination. For example, many pathogenic materials bind to gangliosides, and this has been used as the basis for assays of several toxin materials for close to thirty years. More recently, DNA fragments, called aptamers, have been described, that can potentially be made to recognize specific pathogenic structures. Nucleic acid analysis may also be performed to detect specific nucleic acid sequences that could code for pathogenic markers. Siderophores have also been proposed to be used in this context. Lastly, assays can be performed for products of virulence plasmids, such as the determination of the PA component of *B. anthracis* toxin or some of the various YOP proteins of *Y. pestis*. Examples of various test methodologies are shown in Figure 7.

Parameter	Rationale	Test	Status
Virulence Determination (Many Virulence Factors are contained in plasmid.)	A variety of pathogens and toxins bind to specific receptors such as gangliosides.	Binding to specific receptor	Available (Lab) – Approach has been used with cholera toxin, SEB, and others in a variety of formats
	Many virulence factors associated with bacterial plasmids	Plasmid determinations by Flow cytometry	Demonstrated in laboratory.
		Immunoassay for plasmid products.	Demonstrated in laboratory.
	Binding to a sentinel cell can be measured.	Live cell assays.	Demonstrated in the laboratory.

Figure 7. Methods to determine pathogenicity or virulence factors.

2.4 CLASSIFICATION OF AGENT

The final step in this detection process would be to classify the threat material as a bacteria, virus, proteinaceous toxin, or non-proteinaceous toxin. The most expedient way to make this determination could be through mass spectrometry. Pyrolysis mass-spectroscopy does have this capability, but there are some alternative methods that can be used. Bacteria can be determined through the use of a Gram stain. Specific enzyme activities may also prove useful. For example, beta-galactosidase is an enzyme that is widely used to detect and presumptively identify the presence of *E. coli* in water samples. Some other bacteria also possess similar specific enzymes. Figure 8 shows several approaches for generic bacterial detection/classification.

In the context that these bacteria are used as weapons, one may assume that antibiotic resistance has been introduced into them. There are various tests that are currently available clinically that use a colorigenic substrate that measures an antibiotic lytic enzyme, such as penicillinase.

More recent techniques utilize nucleic acid probes to measure the presence of DNA sequences in plasmids or plasmid constructs that code for these lytic enzymes.

A similar approach can be taken to determine the presence of virus. For example, some viruses possess specific enzyme activities that can be measured. The neuraminidase of Influenza virus is such an example. The properties of this enzyme were studied and were exploited in a rapid assay for detection.

Parameter	Rationale	Test	Status
Bacterial-Specific Fluorescence or Luminescence	Can selectively lyse sample with bacterial specific detergents, enzymes, phages, etc.	ATP Luminescence	Commercially Available
	There is a generic test for E.coli that could be extended to other enterics. (Specific substrates could also be found for other bugs.)	4-MUG Fluorescence	Available – Widely used for water monitoring.

Figure 8. Examples of generic bacterial detection using fluorescent techniques.

Since viruses are intracellular parasites, one may be able to assume that there would be carrier cells or culture components present concurrently with the virus. One might be able to use an assay for ovalbumin in cases where eggs are used as the carrier. If conventional cell culture is used as the means to grow the virus, the mitochondria from them might be measurable by using a fluorescent dye, such as Rhodamine 123. Histocompatibility antigens, which are species-specific antigens that are present on cell surfaces, may also prove useful as a means to detect the presence of viruses. These approaches are summarized in Figure 9.

Parameter	Rationale	Test	Status
Viral-Specific Fluorescence or Luminescence	Neuraminidase is for Influenza detection.	Measure enzymes present on virus.	Demonstrated in the laboratory.
RNA Detection	Most BW viruses contain RNA. RNA content in the environment is low due to lack of active metabolism	Ribo Green Dye fluorescence by flow cytometry or simple fluorometry.	Dye available and demonstrated in the laboratory. Concept demonstrated with marine samples.
Carrier Cells	Viruses would probably have cells available as vectors.	Histocompatibility Antigens by Immunoassay.	Available (Lab) – demonstrated in laboratory.
		Mitochondrial detection using Rhodamine 123.	Available (Lab) – demonstrated in laboratory.

Figure 9. Examples of methods to detect unknown viruses.

Proteinaceous toxins can be determined by several means. Conventional protein determination approaches, such as the Biuret, Coomassie Blue, and others, could serve as an initial screen. This could be followed by more stringent analysis, such as capillary electrophoresis and sequencing. This sequence could then be introduced into a bioinformatics tool, and a possible function could be determined. In the event that the toxin may have some type of enzyme activity associated with it, substrates for the enzyme can be determined and used in subsequent analyses. Figure 10 itemizes some approaches from proteinaceous toxin detection.

Detection of non-proteinaceous toxins would follow a similar approach. Detection could be based on virulence determination, mass-to-charge rations, or enzymatic activity, as shown in Figure 11.

Some of the approaches described in the virulence determination section can also be used to determine if the toxin is dangerous to life and health. For example, the Gm1 ganglioside is found on many cells and many of the pathogenic toxins bind to it. Specific examples include SEB and cholera toxin.

From these approaches, detectors with improved capabilities to detect and warn may be developed that improve one's abilities to protect both troops and assets. Some of these technologies exist today and can be implemented into the field with some success. However, with the improvements in coating technologies and the integration of biological polymers with electronics, the next 50 years should see the development of detectors that utilize several of these approaches in concert.

Parameter	Rationale	Test	Status
Protein Content	Protein content in air is low. Could be used as an additive or agent itself.	Standard Protein Determinations	Available (Field) – Various protection kits are routinely used.
Virulence Determination	A variety of toxins bind to specific cell receptors and modulate the cell's chemistry.	Binding to receptor	Available (Lab) - Approach has been used with cholera toxin, SEB, and others in a variety of formats. Could be integrated into a bio-chip.
	Binding to a sentinel cell can be measured.	Sentinel Cell Assay	Demonstrated in the laboratory.
Marker Determination Sequence Determination	Scope of potential physiological active peptides is great.	Mass Spectrometry	Demonstrated in the laboratory.
Enzymatic Activity	A variety of toxins are either enzymes themselves or inhibit specific enzymatic processes.	Fluorogenic Substrate Analyses.	Demonstrated in the laboratory.

Figure 10. Methods to characterize protein toxins.

Parameter	Rationale	Test	Status
Virulence Determination	A variety of toxins bind to receptors.	Binding to receptor	Available (Lab) – Approach could be integrated into a bio-chip.
	Binding to a sentinel cell can be measured.	Sentinel Cell Assay	Concept demonstrated in the laboratory.
Molecular Weight Determination	Scope of potential physiological active materials is great.	Mass Spectrometry	Demonstrated in the laboratory.
Enzymatic Activity	Substrate Analyses.	A variety of toxins inhibit specific enzymatic processes.	Demonstrated in the laboratory.

Figure 11. Methods to characterize non-proteinaceous toxins.

3. Identification and Medical Countermeasures

3.1 IDENTIFICATION

Once a determination has been made in the field that a biological event has occurred, the next step would be to retrieve the suspect samples and return them to a lab for further identification and classification. There are a plethora of techniques that are available for identification, such as metabolic tests, carbohydrate or fatty acid analysis, phage typing, immunological assays, and nucleic acid probe technologies. Current identification techniques use rigorous analysis to identify microorganisms according to complex taxonomic schemes, using both DNA and RNA analysis. The degree of relatedness among genus, species, and strains can thus be determined. The use of chip technology, although now in its infancy, will play a crucial role in the future in these determinations. Since these technologies are widely known and have been summarized in various reviews, they will not be discussed here. A summary of identification techniques can be found in Figure 12.

Parameter	Rationale	Test	Status
Cellular	Routine test for bacteriological identification	Metabolic fingerprint.	Available (Lab) – Possibly used in field.
	Routine test for strep throat	Immunoassays.	Available (Field)
	Virulence plasmids can be identified by nucleic acid hybridization techniques.	Nucleic Acid Analysis.	Available (Field) – Fieldable nucleic acid analysis system demonstrated on several platforms.
Specification Techniques: Indigenous vs non-indigenous strains	Accepted techniques for bacterial identification	Chromosomal DNA analysis	Available (Lab) – can be implemented into field situations.
		RNA fingerprinting	Available (Lab) – can be implemented into field situations.
Genetic Manipulation	Identify vector sequences. Identify markers associated with environmental stability.	Nucleic analysis	Available (Lab) – Needs additional development to expand available library.

Figure 12. Methods for Identification of biological agents.

3.2 MEDICAL COUNTERMEASURES

3.2.1 Rapid Susceptibility Testing

A current rationale, that one needs to know the identity of the particular agent so that the proper treatment modality can be employed, needs to be re-evaluated. In classical medical approaches, when one knows the identity of the organism, one can typically prescribe the appropriate course of antibiotics or other therapies. However, with the advent of genetic engineering and the relative ease that this allows an adversary to impart resistance to antibiotics, one can no longer assume that the mere identification of the organism would be sufficient for treatment. Even in conventional medical treatment, there has been an increase in the use of susceptibility tests to determine the appropriate course of treatment. In the case of an intentional release of a biological agent by an adversary, the use of antibiotic susceptibility testing is crucial.

In the event that a biological attack has occurred, particularly with a bacteria or a toxin, a viable sample is crucial for initiation of medical countermeasures. Although Koch's postulates will have to be demonstrated for legal purposes, the viable sample will be necessary so that antibiotic susceptibility testing can be performed. Currently these tests involve isolating the organism and eventually obtaining it in pure culture for further analysis. In the case of a bacterium, the Kirby-Bauer procedure is usually used and takes 8 or more hours to complete. In the event of a biological attack, there may not be sufficient time to do this. What is needed is a rapid method to do these determinations on environmental samples.

Several approaches have been proposed to approach this. One utilizes flow cytometry and rapid determinations in 20 minutes have been achieved. This approach uses a classical approach where the viable cells are mixed with an antibiotic mixture. The effect of the antibiotic on the cells is then measured by changes in scatter or DNA-specific dye uptake. Alternatively, analysis with nucleic acid probes that are specific for sequences that code for antibiotic resistancehave been proposed. This has the added advantage over conventional techniques in that viable samples are not required.

A similar paradigm can be assumed for viruses, but toxins are a different case. Specific identification needs to be obtained so that the proper antidotes, if available, can be administered. In the case of real unknown entities, bio-informatics will eventually play a role so that the possible physiological activity of the material can be obtained. However, we are several years away from this being a field consideration.

3.2.2. Immune Status Determinations

Up until this point, the agents themselves have been discussed; however, the other part of the equation concerns the troops that may have been attacked. It may be desirous to make a determination as to who has been attacked so that the field commander may not have to compromise his military posture. One possibility is to measure the status of immune function of the individuals who were involved in the attack. There are several approaches that may be used to assess immune status, from the classical blood cell counts to measuring individual lymphocyte status by flow cytometry analysis. As our ability to measure these functions improves, they may prove to be viable assets in the field.

4. Methodology for Optimum Equipment Determination

So how does one determine the best equipment for the task at hand? The first step is to develop a response plan and determine how the response will be executed and who will be responsible for what actions. Ultimately one needs to determine who the end user will be; how they operate; the operational constraints; the performance criteria for the equipment; and what are the impacts of each of these parameters on the overall goal. Put another way, one needs to determine the weighting factors of the various personnel, operational, and performance constraints.

Once these factors are identified, they can then be used with a decision-tool/aid software package to assist in the decision making process. There are a variety of these programs available and this would be an ideal application for them.

We recently undertook such an effort. A panel of experts was convened and the operational parameters were discussed. It was decided that the end user required both a generic detection system and a field portable identification system that was instrumented. The operational constraints did not allow culture as a possible solution. A model for both of these scenarios was then established and the various parameters and weighting factors were built in. Various types of detectors were chosen and characterized according to the parameters established by the model. Detectors included relatively screening tests as described above to more complex systems that utilized expensive equipment. The data from this characterization was then entered into the model and evaluated by the algorithm. An example of the detector model is given in Figure 13. Data for the model is given in Figure 14 and the subsequent output is given in Figure 15.

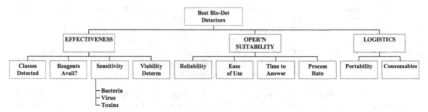

Figure 13. A decision model for the selection of field deployable biological screening (detector) devices.

	Classes Detected 0.20	Reagents Avail? 0.05	Bacteria Sens 0.05	Virus Sens 0.05	Toxins 0.05	Viability Deter 0.20	Reliability 0.15	Ease of Use 0.15	Time first answer 0.05	Process Rate 0.02	Portability 0.02	Consumables 0.01
Detector 1	Four	100	100,000	10,000K	100	100	90	60	80	20 or more	Man port	Low cost
Detector 2	Three	65	1,000	10,000K	0	75	70	100	100	20 or more	Man port	Low cost
Detector 3	Three	65	1,000	None	0	75	70	100	100	20 or more	Man port	Low cost

Figure 14. Example of input data for three hypothetical detector systems.

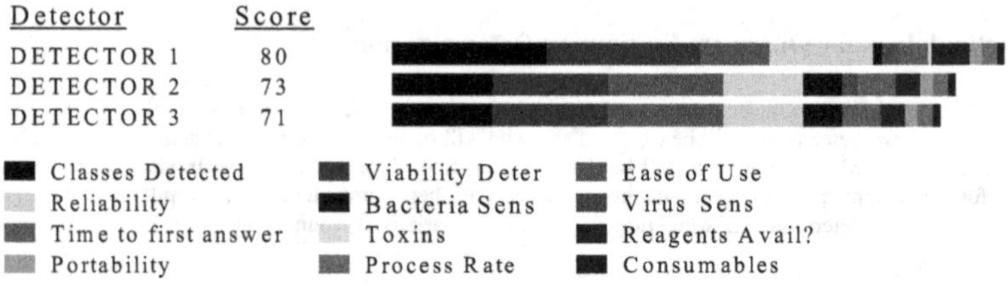

Detector	Score
DETECTOR 1	80
DETECTOR 2	73
DETECTOR 3	71

- Classes Detected
- Reliability
- Time to first answer
- Portability
- Viability Deter
- Bacteria Sens
- Toxins
- Process Rate
- Ease of Use
- Virus Sens
- Reagents Avail?
- Consumables

Figure 15. Output from a hypothetical detector model.

A model for an identifier system is given in Figure 16 and examples of output data from the model is shown in Figure 17. IA denotes immunoassay devices while PCR denotes nucleic acid analysis devices.

Figure 16. A Model for an Identifier System

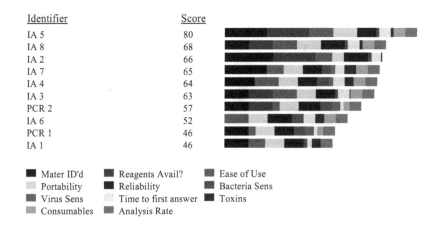

Figure 17. Output from an identifier model.

The output of this model shows that a particular immunoassay format is perhaps the best choice for the particular effort, mainly due to the extent of the material that is identified. The next several immunoassay devices scored reasonably close together, although the individual differences that resulted in this clustering are different. One cannot really say that the assays are similar because of the differences in the other operational parameters. Perhaps a re-evaluation of the weighting factors could help to separate this middle group if the highest scoring system was not available. The PCR system scored lower than the immunoassay systems, the driver being the then level of maturity of these systems and the time to first answer, along with the analysis rate.

Biological agents, on the other hand, are a completely different story. Technologies for BWA are largely derived from the clinical side and still need to be developed into a fieldable system. There are systems that can be utilized for rapid screening, but the applications of these technologies need to be accepted by all parties.

5. Deployment Strategies

The previous discussion dealt with the types of technologies that one could conceivably use to detect the presence of biological materials; however, concepts of employment still need to be determined. This is the difficult part because it is here where the considerations are more cost/benefit, logistics, and personnel driven, rather than science driven. If we were to determine the types of scenarios that one would get in a potential terrorist scenario, there would be 2 cases: high value, fixed assets (buildings, stadiums, etc.) and hoax scenarios. In both of these cases, the scenarios are quite different. Another consideration is responding post attack. Here we need to determine of areas are contaminated and then perform quality control on personnel and materiel decontamination.

If there are some high value, fixed assets where a threat is credible, then some type of continuous monitoring system is probably worthwhile considering. These types of systems could utilize one or more of the triggering/detection technologies. For example, particle size and shape analysis coupled with fluorescence, could be an effective system. Systems such as the CDC 4WARN or the BAWS system under development in the US could be likely candidates. To minimize false alarms, this approach could be coupled with a detector system that utilizes some of the detector schemes, such as bio/non-bio or pathogen/non-pathogen. An ideal candidate for this type of approach is a flow cytometer or similar device whereby several parameters can be coupled together in one platform. However, the down side of this approach is that it would require some degree of maintenance by building personnel.

Hoax scenarios can be dealt with by relatively simple equipment. One initially needs to make a bio/non-bio determination by a DNA or protein test. If positive, a viability test could then follow to determine if the sample contains live organisms. Samples can then be further analyzed by identification technologies, either on or off site.

Responding to an attack, however, requires an echelon of response. One first needs to determine if there is indeed a biological agent present. This can be determined by the same means discussed in the hoax scenario. If it is determined that indeed live biological material is used, then the area can be sampled to determine the extent of contamination. Lastly, the area can then be decontaminated and the effectiveness of this process can be determined by several simple tests, such as luminescence.

6. Conclusions

From this brief discussion, it can be shown that the problem is not insurmountable; however, several things need to change. We first need to become aware that the possibility does indeed exist that "unconventional" biological agents be encountered in the field. There are a variety of strategies that could be implemented in either trigger or detection platforms that could be used to detect signatures from biological agents and possibly determine that they could present a danger to health and life.

There are some software decision tools that can aid in the selection of CWA and BWA sampling and detection equipment. Requirements and weighting factors need to be established before technologies can be evaluated by the model. These models can be useful in the selection of responder equipment for CWA and BWA responses.

However, we need to change our paradigms on how we think about the problem. The most important thing is to provide a detection and warning system that exploits credible signatures so that those in peril can take the appropriate protective measures.

7. Bibliography

Adam P. Flame photometry for biological detection. Proceedings of the Sixth International Symposium on Protection Against Chemical and Biological Warfare Agents, 1998, 61-67.

Bruno JG, Mayo MW. A color image analysis method for assessment of germination based on differential fluorescence staining of bacterial spores and vegetative cells using acridine orange. Biotechnic and Histochemistry 1995; 70(4):175-184.

Bruno JG, Yu H, et al. Development of an immunomagnetic assay system for rapid detection of bacteria and leukocytes in body fluids. J. Mol Recog 1996; 9:474-479.

Bruno JG, Kiel JL. In vitro selection of DNA aptamers to anthrax spores with electroluminescence detection. Biosensors and Bioelectronics 1999;

Button DK, Robertson BR. Determination of DNA content in aquatic bacteria by flow cytometry. Appl. Envrion. Microbiol. 2001; 67(4): 1636-1645.

Darzynkiewicz Z, Bedner E, et al. Laser-Scanning cytometry: a new instrumentation with many applications. Experimental Cell Research 1999; 249:1-12.

Davey HM, Kell DB. Flow cytometry and cell sorting of heterogeneous microbial populations. Microbiol. Reviews 1996; 60:641-696.

Ekins RP. Immunoassay, DNA analysis, and other ligand binding assay techniques: From electropherograms to multiplexed, ultrasensitive microarrays on a chip. J Chem Ed 1999: 76(6): 769-788.

Ezzell JW, Abshire TG. Immunological analysis of cell associated antigens of *Bacillus anthracis*. Infect. Immun. 1988; 56:349-356

Ezzell JH, Abshire TG, et al. Identification of *Bacillus anthracis* by using monoclonal antibody to cell wall galactose-N-acetylglucosamine polysaccharide. J Clin Microbiol 1990;28:223-31.

Forsberg A. A shared strategy for virulence of bacterial pathogens. Proceedings of the Sixth International Symposium on Protection Against Chemical and Biological Warfare Agents, 1998, 163-172.

Hairston PP, Ho J, Quant FR. Design of an instrument for real-time detection of bioaerosols using simultaneous measurement of particle aerodynamic size and intrinsic fluorescence. J. Aerosol Sci. 1997; 28(3): 471-482.

LeBaron P, Servais P, Agogue H, Courties C, Joux F. Does the nucleic acid content of individual bacterial cells allow us to discriminate between active cells and inactive cells in aquatic systems? Appl. Environ. Microbiol. 2001;67(4): 1775-1782.

Mason DJ, Shanmuganathan S, et al. A fluorescent gram stain for flow cytometry and epifluorescence microscopy. Appl environ Microbiol 1998: 64(7): 2681-2685.

Maltsev VP, Cheruysvev V. Method and device for determination of parameters of individual microparticles. US Patent Number 5,560, 847, issued 22 July 1997.

Olive DM, Bean P. Principles and applications of methods for DNA-based typing of microbial organisms. J Clin Microbiol 1999; 37(6):1661-1669.

Robertson BR, Button DK, Koch AL. Determination of the biomasses of small bacteria at low concentrations in a mixture of species with forward light scatter measurements by flow cytometry. Appl Environ Microbiol 1998; 64(10): 3900-3909.

Rolland X. Chemscan™ RDI: A real time and ultra-sensitive laser scanning cytometer for microbiology. Applications to water, air, surface, and personnel monitoring. Proceedings of the Sixth International Symposium on Protection Against Chemical and Biological Warfare Agents, 1998, 103-110.

Rowe CA, Tender LM, et al. Array biosensor for simultaneous identification of bacterial, viral, and protein analytes. Analytical Chemistry 1999: 71(17); 3846-3852.

Sincock SA, Kulaga H, et al. Applications of flow cytometry for the detection and characterization of biological aerosols. Field Anal Chem Tech 1999:3:291-306.

Stopa PJ Tieman D, et al. Detection of biological aerosols by luminescence techniques. Field Anal Chem Tech 1999:3:283-290.

Stopa PJ and Bartoszcze MA. Rapid Methods for Analysis of Biological Materials in the Environment. NATO ASI Series, KluwerAcademic Publishers, Dordrecht, NL, 2000.
Some specific articles of interest include:
    Bartoszcze M, Bielawska A. The Past, Present, and Future of Luminometric Methods in Biological Detection, 73-78.
    Boulet CA, Hung G, et al. Capillary Electrophoresis/Nucleic Acid Probe identification of Biological Warfare Agent Simulants, 87-92.
    Bryden WA, Benson RC, et al. Tiny-TOF Spectrometer for BioDetection, 101-110.
    Ho J, Spence M, and Hairston P. Measurement of Biological Aerosol with a Fluorescent Aerodynamic Particle Sizer (FLAPS): Correlation of Optical Data with Biological Data, 177-201
    Del Vecchio VG, Redkar R, et al. Development of PCR-Based assays for the detection and molecular genotyping of microorganisms of importance in biological warfare, 219-229.
    Garrigue H, Patra G, and Ramisse V. Use of PCR for Identification and Detction of Biological Agents, 259-278.

Stopa PJ. The flow cytometry of *Bacillus anthracis* spores revisited. Cytometry 2000; 41(4): 237-244.

Walberg M, Gaustad P. Steen HB. Rapid assessment of ceftazidine, ciprofloxacin, and gentamicin susceptibility in exponentially-growing *E. coli* cells by means of flow cytometry. Cytometry 1997;27:169-178.

Walt DR, Franz DR. Biological Warfare Detection. Analytical Chemistry 2000; December 738A-746A.

Wiener SL. Strategies for the prevention of a successful biological warfare aerosol attack. Military Medicine 1996; 161(5): 251-256.

Zubkov MV, Fuchs BM, et al. Determination of total protein content of bacterial cells by SYPRO staining and flow cytometry. Appl Environ Microbiol 1999; 65(7):3251-3257.

ATP BIOLUMINESCENCE FOR THE DETECTION AND IDENTIFICATION OF BIOLOGICAL THREATS

MICHAL BARTOSZCZE AND AGATA BIELAWSKA-DROZD
Military Institute of Hygiene and Epidemiology,
Lubelska 2, 24-100 Pulawy, Poland

1. Introduction

Bioluminescence is a known method that has been used from many years in the pharmaceutical and brewery industries [12]. Wider introduction of this method into the practice was limited by time-consuming procedures, imperfect equipment, reagent instability etc. The progress in the technology of photon detectors, electronic circuits, and reagents has resulted in renewed interest in bioluminescence. NASA exhibited attention to the benefits of this technique during their search for life efforts. Private companies have also found the advantages of this technique and have implemented them in their production processes, largely in quality control of hygiene. So the re-emergence of bioluminescence is connected with, among others, its high sensitivity, near real time or rapid analyses, and point of interest analysis, which is away from the laboratory. The versatility of bioluminescence application is worthy of attention, making it one of the more useful analytical methods.

2. Bioluminescence Basics

The word "bioluminescence" derives its origins from the Greek word *bios* (life) and the Latin word *lumen* (light). The bioluminescence phenomenon is observed in the environment in some species of bacteria, e.g. Vibrio (*V. harveyi, V. fischeri*) and Photobacterium (*P. phosphoreum, P. liognathi*). It also is observed in fungi, as a by-product of metabolism processes; in some protozoa; insects (*Photinus pyralis* – the firefly); coelenterates; molluscs; crustaceans; and fish.

2.1 REACTION MECHANISM

The oxidation of luciferin by the luciferase enzyme is the basis of the bioluminescence phenomenon. Oxy-luciferin is initially created by the enzyme-substrate reaction. The activated state then causes the reduction of high-energy ATP (adenosine tri-phosphate) to AMP (adenosine mono-phosphate) to the basic form, the result of which is the emission of photons. The amount of these photons is proportional to the ATP content in the tested material [29]. The bioluminescence technique allows the direct measurement of photon emission in the tested material containing ATP. Because ATP is a basic compound occurring in living cells (except viruses) and is required for "life", its detection indicates the presence of "live or viable" cells that originate from human or animal tissues, plants, and micro-organisms. The presence of ATP is, therefore, an indicator of biological contamination.

The reaction scheme was presented below:

LUCIFERIN + LUCIFERASE + ATP -----Mg^{2+} -$\rightarrow$> LUCIFERIN +PPi (pyrophosphate) + LUCIFERASE— MAP

LUCIFERIN + LUCIFERASE—MAP -----O_2 ------$\rightarrow$> OXYLUCIFERIN + LUCIFERASE + CO2 + AMP + LIGHT (625 nm).

The bioluminescence reaction occurs extremely fast. The reaction is initiated when the luciferin/ luciferase/ Mg^{2+} substrate mixture is introduced into the test tube containing the sample. If ATP is present in the sample, the reaction is initiated and an intense photon emission occurs. This is followed by a gradual reaction inhibition that is caused by

P. Stopa and Z. Orahovec (eds.), Technology for Combating WMD Terrorism, 67-74.
© 2004 *Kluwer Academic Publishers. Printed in the Netherlands.*

the increase of the concentration of final product {Ppi-(pyrophosphate)}. The bioluminescence reaction kinetic curve shows two-phase light activity.

2.2 LEVELS IN CELLS

The ATP level differs depending on the type of material examined. In bacterial cells, it amounts to about 1 pg ATP [9], and varies between 0.1 and about 5 pg. This depends on the type of the bacteria and its metabolic state. The ATP level in single bacterial cells has values from several to tens of mmol/mg per dry mass. For instance they amount: for *E. coli* 3.57, *B. cereus* 7.7, *Beneckea harveyi* 13 – 16, and K. *aerogenes* 3.7 – 8.3 [9]. On the other hand, yeast cells and mammalian cells contain a similar amount of ATP, e.g. about 10 – 12 pg [11].

The luciferase isolated from *Photinus pyralis* (glow worm) is the enzyme used most often in bioluminometry. It belongs to the oxygen dehydrogenase of flavoprotein group, catalyzing reactions of oxidative decarboxylation of luciferin in the presence of Mg $^{2+}$ ions. The molecular weight of the enzyme is about 100 kilodaltons, resulting from the presence of two identical fragments with the molecular mass 50 kilodaltons [29].

2.3 FACTORS THAT INFLUENCE THE COURSE OF THE REACTION

There are different agents that influence the course of the enzymatic reaction, such as temperature (optimum 20 – 25 ^{o}C) and pH (optimum=7.8). When the reaction occurs at the optimum parameters, the wavelength of emitted yellow–green light is observed at 562 nm .

Conversely, false negative results may be caused by thepresence of heavy metal ions, detergents, disinfecting agents, antibiotics, ADP, and other factors that may be present in the samples [33,35,36]. Some chemical substances my act synergistically, causing the photon emission to increase. These observations are very important in a practical sense because they may introduce significant errors in luminometric determinations, influencing the proper interpretation of results.

Some of these factors can be mitigated by the selection of appropriate buffers. In most cases, however, it is necessary to use modern sample preparation technologies to eliminate the interfering agents influence on the assay result.

3. Luminometry Applications

3.1 DETECTION OF BACTERIAL CONTAMINATION ON SURFACES

Rapid detection of a biological agents presence is extremely important for the protection of human health and life. The detection of bacterial contamination on a surface allows one to execute immediate intervention, without the necessity of a long wait for the results made with traditional methods (culture), taking usually 24 – 72 hours. In practice, one is chiefly concerned with surfaces that are in contact with the process of preparation, production, processing, storage, and serving of food. A high level of bacterial contamination on surfaces indicates the potential presence of pathogenic micro-organisms, such as *E. coli* and *Salmonella spp.* These surfaces include transport devices, cloths, personnel's hands, dishes [4].

The luminescence method may be successfully applied for critical point monitoring in the Hazard Analysis of Critical Points (HACCP) system, as well as potential sources of contamination detection in production plants [21]. The necessity of the differentiation of somatic and bacterial cells is the basic problem in using this technique for the examination of surfaces. Somatic cells possess 10 – 100 times more ATP when compared with bacterial cells. The results obtained, without the differentiation of bacterial and somatic contamination, probably show a false picture of the state of hygiene in the process.

In the evaluation of surface contamination, the presence of interfering substances in samples must be absolutely eliminated. The presence of the previously described interfering agents that may inhibit the bioluminescence reaction may incorrectly indicate the lack of cellular (bacterial or somatic) ATP presence, and thus lead one to conclude there is a lack of bacteria present on the surface. Such a situation could result in a large hazard, one that has large consequences on human health (food poisoning) as well as economical ones.

Another factor that may influence the results of surface determinations is the type of swab used for sample collection [23]. A previous study has shown that swabs made of cellulose fibers may inactivate bacteria because of the presence of sulphur compounds. Another study showed that some types of fibers bind bacteria strongly, making it difficult to recover them for testing. However, Dacron swab proved to be completely suitable, giving the best sample recovery [31].

The collection of surface samples is executed with the use of a suitable swab and the appropriate sampling technique. Typically a pattern is used whereby one swabs a surface, using both horizontal and vertical moves. The exact pattern is determined by local protocol or requirements. A 10 cm x 10 cm pattern is the norm. If the surface to be sampled is dry, the swab should be wetted with the use of a sterile solution of physiological salt. If the tested surface is wet, however, a dry swab can be used. The swab is then put into the test tube containing the physiological fluid. One can then use rotating moves and pressure against the tube wall to extract the sample. The analysis is then performed according to the following procedure.

The Profile-1™ luminescence system, manufactured by New Horizons Diagnostics, Inc, Columbia, MD, USA, significantly removes those factors that may have an adverse effect on the luminescence reaction. This system uses a device, called a "filtravette", which allows one to both filter and selectively treat (lyse) interfering substances. The ATP of somatic cells is first selectively lysed and then filtered into an absorbing paper. The filter, however, efficiently captures untouched bacterial cells on its surface. In this way the ATP from somatic cells, along with potential inhibitors, such as kitchen salt, disinfecting agents (detergents), heavy metal ions etc., are removed. Bacterial cells remaining on the filter surface are then treated with a reagent that liberates the only the bacterial cell ATP. The luciferin / luciferase (L/L) reagent is subsequently introduced. If the cells are alive, the presence of ATP initiates the luminescence (light) reaction. The emitted light intensity is determined in an apparatus that is equipped with a sensitive photo-detector. This device provides a result, given in Relative Light Units (RLU) that one can use to correlate to bacterial concentration.

For the determination of acceptable and unacceptable bacterial contamination levels, a control surface should be chosen that can be disinfected. Results obtained from this will be treated as the reference point for further luminometric determinations, the results of which may be the control point in HACCP system. A study by Hasan et al.[13] has shown a 90% correlation of results obtained from contaminated surfaces when luminescence and traditional culture are compared. A study conducted by the Canadian Food Inspection Agency showed a 100% correlation between luminescence and culture for 96 surface samples collected via the swab method in food producing plants. In another study [5], the luminescence method exhibited sensitivity for surfaces contaminated with $E.coli$ of 40 CFU/cm^2. Deininger and Loomis obtained similar results [9], and also demonstrated sensitivity toward fungi of 10 CFU/cm^2.

3.2 ANIMAL CARCASSES AND FOOD

The luminometric technique was used successfully for the assessment of biological contamination of cattle, pork, poultry carcasses, fresh meat and raw milk [1,6,14,15,24,27,28,34]. Carcasses often undergo fecal contamination during slaughter with negative impact on food safety. Feces are the source of many dangerous bacteria. Contaminating bacteria include *Salmonella* spp., *Campylobacter jejuni*, *Escherichia coli* (O157:H7), and other bacteria, commonly referred to as fecal coliforms. Traditional methods for the determination of carcass contamination are based on the culture methods. These techniques are time–consuming and require significant training to both execute and interpret the results. Therefore, there is no possibility to make immediate decisions on the status of the production process. The application of the luminometric method does allow one to make immediate decisions for quality contraol and/or assurance on the process because results are obtained in near real-time. Studies of cattle and pork carcasses conducted by the USDA showed a high correlation (r = 0,91 – 0,93) between luminometry and culture when evaluating these carcasses for contamination. The authors recommended the use luminometry method for production process monitoring according to the HACCP program [27,28].

3.3 WATER TESTING

Several studies found that the luminometry method was suitable for rapid testing of biological contamination in water [5,7,8,9,10,16,32]. Deininger et al [8,9] performed a study where he collected and tested on-site water samples from different parts of the world. His results showed a high degree of correlation (r = 0,99) between the results of the luminometric method and the standard culture method (Heterotrophic Plate Counts – HETP). Thus, the luminometric method can be used for "water safety" monitoring; can be used to perform quality control measurements on water conditioning systems; and can be used to rapidly detect intentional contamination. The rapid performance and execution of the test)(10–15 minutes) and the method's sensitivity (10– 40 CFU/ml) suggest that it would make an excellent field test for this purpose. The study of the Paris water supply system made by Delahaye et al [10] indicates the suitability of luminometry for "alert" monitoring of water contamination with biological threats.

3.4 AIR MONITORING

Luminometry may be used successfully for monitoring of air for the presence of high levels of bacteria that may suggest the presence of contamination. These tests are very important in hospital management, pharmaceutical plants, food production plants, cosmetics plants, etc. A sensitivity of about 300 CFU was demonstrated [7], which depends upon the sample collection technique and collected volume of the air. According to Stopa et al [30], the ATP determination sensitivity in the air amounts 100 pg/ml and has very good correlation with bacteria culture methods.

3.5 POWDER

3.5.1 Examination for the presence of bacterial spores.
In cases where the surface contamination may be a powder, samples can be collected by either a swab or a sponge method. The luminometer can then be used to detect vegetative bacteria by the conventional method; however, a separate luminometric test should also be performed for the detection of spores present in samples. The luminometric detection of vegetative bacterial ATP presents no difficulties, but this is not the case for spore detection. The complicated structure of the spore cell wall, the difficulties of liberating ATP, and the small content of ATP are the crucial difficulties. We developed a method [2] whereby thermal shock is first used for *Bacillus anthracis* spores, and then transferred to a both that is enriched with L- alanine and adenosine to initiate the germination process. The spores are then incubated at a temperature of 37 °C. Observable ATP generation during spore germination was observed after 5 minutes. Stopa et al [30,31] obtained similar results, with the omission of the thermal shock and with the use of Trypticase Soy Broth (TSB). Either method allows for rapid spore detection and may be suitable for to study the germination process itself. These techniques have been used in screening studies of large amounts of samples suspected of containing *B.anthracis* spores. According to Deininger et al [7], the luminometry can detect approximately 300 spores.

3.5.2 Other Applications
The high sensitivity of the method; its simplicity; and the rapidity with which the test can be performed show that there is a large future in front of the luminometry. It is used, for example, for quality control and improvement of the state of sanitation and hygiene in health care institutions [26], meat processing plants, poultry processing plants, dairy plants, fruit and vegetable processing plants, distilling plants, winemaking, sugar industry, and animal breeding. Other applications include the monitoring of water treatment stations and mineral water production lines for quality control regimes for microbiological contamination. This technique has application to other areas of production line monitoring, including raw material, semi-completed products, and finished products admittance points. For food applications, this technique may be successfully applied to the whole production process by implementation into the HACCP system (critical control points). We have also used this technique to in the study of the efficiency of disinfecting agents [3], as well as for water disinfecting efficiency assessment purposes. These are additional examples of the versatility of luminometry.

Nowadays, ATP bio-luminescence is used widely for biomass determination in limnology, oceanography, and to the study of biological activity of sludge (mud) whereby one determines the amount of micro-organism ATP content during their growth processes [29] . The luminometry method can also be used to determine the bacterial contamination of fuel [22].

The luminometry method does not allow one to detect presence the presence of viruses since they do not contain ATP. Lately, however, it was shown on the Aujeszki virus model, that these the implementation of luciferase gene to the virus genome can provide the virologist with an additional tool. The light reaction can be detected on photosensitive film. A similar study was performed with the use of a cDNA copy that was introduced into the cowpox virus genome.

4. Identification of Biological Agents with Luminometry

4.1 IMMUNOBEADS

The classical luminometric technique for the detection of bacterial ATP is a non-specific method since it does not allow for bacterial differentiation. The real difficulty is to find a way to combine bioluminescence with a micro-organism identification method. We conducted a study in our laboratory [19], using *Salmonella* as a model, whereby we covered paramagnetic beads with specific antibodies, allowing the selective "catching" of bacteria from the suspension. This made their identification possible with the use of luminometric technique was. This procedure is rather simple, and the time to results is extraordinarily short comparing with the classic method. The paramagnetic beads (PMB) application immobilized with apyraze significantly improved the luminometric reaction sensitivity . Recently, Lee and Deininger [16] used a similar method for detection and identification of *E. coli*(0157:H7) in water. They used paramagnetic beads that were covered with specific antibodies that enabled them to detect the bacteria efficiently and quickly. They also indicated the possibility of the application of the method for the detection of bioterrorist agents in water.

4.2 MOLECULAR BIOLOGY

The application of the luciferase gene in molecular biology [17,18] is very promising. The gene responsible for bacteria bioluminescence was identified and cloned. The DNA fragment containing this gene was then introduced into the nucleic acid of a phage that is specific for the target cell. Once the bacterial cell is infected, the phage replication occurs, and the transcription process is accompanied with light emission. The P-22 phage, which is specific for *Salmonella typhimurum*, was used to show the possibility of detecting 100 bacteria within 30minutes. This work is very promising, especially since it may provide a means of rapid identification of the agents that cause food poisoning.

4.3 PHAGE ENZYMES

Enzymes obtained from bacteriophages have recently been utilized with luminescence techniques. These enzymes have unique features, where specificity for the target cell is the most important one. A study conducted at the Rockerfeller University in the USA [20,25] demonstrated that the lytic enzyme obtained from the gamma phage lysed, and then cleared, the cell wall of *B. anthracis*. This enzyme was then applied to luminometry where it liberated the ATP from vegetative *B. anthracis*, whereby the detected ATP indicates this bacteria presence in the sample. This phage enzyme phage enzyme specifically destroys *B. anthracis*, and showed no lytic activity toward other bacteria , e.g. *B. subtilis, B. cereus, B. thuringiensis* etc.

Similar specific activity was also demonstrated with a phage enzyme specific for Group A Streptococci. This study prepared the way for a wide range of application of these enzymes to other micro-organisms, thus bringing significant progress for their rapid identification and new hope for the fight against infections, too (25).

5. Conclusions

The discussion presented here shows that ATP luminescence is a valuable detection method of the future. It can be successfully implemented for the detection and identification of biological threats in the field and in the laboratory, thus improving our ability to rapidly detect and identify bacteria.

6. References

1. Arciuch H., Bartoszcze M., Trudil D., Matras J.;(1996) Luminometry of bacterial ATP in milk. Proc. of Fourth ASEPT International Conference. Securite Alimentare 96, Food safety 96, 4-6 June , Laval, France, 356.

2. Bartoszcze, M., Arciuch, H., Chomiczewski, K., and Matras, J. (1997) Some problems concerning application of the luminometric methods in the detection of *Bacillus anthracis* spores. Proc. of the 1996 ERDEC Conference on Chemical and Biological Defense Research, 19-22 November 1996, Aberdeen Proving Ground, 711-712.

3. Bartoszcze, M., Bielawska, A., Matras, J., and Chomiczewski, K. (1998) Application of bioluminescence in the rapid detection of active *Bacillus anthracis* spore inhibitors. Proceedings of the 1997 ERDEC Scientific Conference on Chemical and Biological Defense Research, 18-21 November 1997: 611-612.

4. Bartoszcze, M., Chomiczewski, K., and Bielawska, A. (1998) Bioluminescent technique in rapid microbiological detection. XXXIInd International Congress on Military Medicine, 19-24 April Vienna, Austria.

5. Bartoszcze, M., Chomiczewski, K., Bielawska, A., Mizak, L., and Szymajda U. (1999) Bioluminescence sensitivity regarding microbiological surface and water contamination. Proc. of the 1998 ERDEC Scientific Conference on Chemical and Biological Defense Research, 17-20 November 1998: 525-527.

6. Cutter, C.N., Dorsa, W.J., and Siragusa G.R. (1996) A rapid microbial ATP bioluminescence assay for meat carcasses. Dairy Food & Environ. San. **16**:726-736.

7. Deininger R.A., Lee J.: Estimation of viable spores in powder, air and water samples in 5 minutes.(rad@umich.edu)

8. Deininger, R.A., and Lee J. (2001) Rapid Determination of Bacteria in Drinking Water Using an ATP Assay. Field Analytical Chemistry and Technology **5**:1-5.

9. Deininger, R.A., and Loomis, L. (2001) Rapid Detection of bacteria in water, air and on surfaces. Luminescence Information Packet. New Horizons Diagnostic Corporation

10. Delahaye, E., Welte, B., Levi, Y., Leblon, G., and Montiel, A. (2003) An ATP-based method for monitoring the microbiological drinking water quality in a distribution network. Water Res. **37**:3689-3696.

11. Garg, S., et al. (1998) Rapid enumeration of yeast ATP by bioluminescence. 98th Gen. Meet. Am. Soc. Microbiol. Abst. No **13**:416.

12. Hysert D.W., Kovecses F., Morrison (1976) A firefly Bioluminescence ATP Assay Method for rapid detection and enumeration of brewery microorganisms. J.Am.Soc.Chem.34:145-150

13. Hasan, J.A.K., Garg, S., Loomis, L., Miller, D., and Cowell, R.R. (1997)Direct detection of bacteria from surfaces in 5 minutes with 90% correlation to standard culture. 97th General Meeting of Am. Soc. Microbiol. Abst. No. I:126.

14. Kennedy J.E., Oblinger J.L.: (1985) Application of Bioluminescence to Rapid Determination of Microbial Levels in Ground Beef. J. Food Prot.48: 334-340

15. Kenneth J.L.,Pikelis S.,Spurgash A.(1986). Bioluminescent ATP Assay for Rapid Estimation of Microbial Numbers in Fresh Meat. J.Food Prot.49:18-22

16. Lee J., Deininger R.A. A practical method to detect Biological warfare agents in potable water.(2000) Conference on Bioterrorism and Biodefence, Am. Society of Microbiology, Lansing, MI

17. Lee S., Masayasu, S., Tamiya E., and Karube, J. (1991) Microbial detection of toxic compounds utilizing recombinant DNA technology and bioluminescence. Analytica Chimica Acta **244**:201-206.

18. Lee S., Masayasu S., Tamiya E., and Karube J. (1992) Sensitive bioluminescent detection of pesticides utilizing a membrane mutant of *Escherichia coli* and recombinant DNA technology. Analytica Chimica Acta **257**:183-189.

19. Lidacki A., Bartoszcze M., Arciuch H., Skoczek, A., and Mierzejewski, J. (1995) The evaluation of IMS method for biological detection. Proc. of the ERDEC Scientific Conference on Chemical and Biological Defense Research, 14-17 November 1995: 775-777

20. Nelson, D., Loomis, L., and Fischetti, D.A.Using bacteriophage lytic enzymes as a diagnostic tool for rapid identification of specific bacteria. The Rockefeller University C-40, 1-14

21. Niedoba, E., and Jackowski, D. (1999) Praktyczne wykorzystanie bioluminometrii w pracy wojskowego inspektora weterynaryjnego. Życie Weterynaryjne **74**: 207

22. Passman F.J., Loomis L., Sloyer J.(2003).Non-conventional methods for estimating fuel system bioburdens. IASH 2003, The 8th International Conference on Stability and Handling of Liquid Fuels Steamboat Springs, Colorado, September 14-19

23. Perry J. L., Ballou, D.R., and Salyer J.L. (1997) Inhibitory properties of a swab transport device. J.Clin. Microbiol. **35**:3367-3368.

24. Quessy S., Letellier A., Roseberry K.(1997) Monitoring of fecal contamination of swine carcasses using a portable bioluminometer. World Congress on Food Hygiene, The Hague, The Netherlands, 308.

25. Schuch, R., Nelson,D., Fischetti,V.A.(2002) A Bacteriolytic agent that detects and kills Bacillus anthracis. Nature , 418: 884-889

26. Seeger K.,G., Griffiths K.S. 1994, Adenosine Triphosphate Bioluminescence for Hygiene Monitoring in Health Care Institutions. J.Food Prot.57: 509-512

27. Siragusa, G.R., Cutter C.N., Dorse, W.J., and Koohmaraie, M. (1995) Use of a rapid microbial ATP bioluminescence assay to detect microbial contamination on beef and pork carcasses. J. Food Protect. **58**:770 - 775.

28. Siragusa, G.R., Dorsa, W.J., Cutter, C.N., Perino, L.J., and Koohmaraie, M. (1996) Use of newly developed rapid microbial ATP bioluminescence assay to detect microbial contamination on poultry carcasses. J. Biolumin. Chemilumin. 11:297- 301.

29. Steinberg S.M., Poziomek E., Engelmann W.H., Rogers K.M.: A review of environmental application of Bioluminescence measurements. (1995) Chemosphere 30: 2155-2195

30. Stopa, P.J, Tieman, D., Coon, F.A., Milton, M.M., and Paterno, D. (1999) Detection of biological aerosol by Luminescence Techniques. Field Analytical Chemistry and Technology 3:283.

31. Stopa, P.J., Coon, P.A., Seitzinger, A.T., and Paterno, D. (1998) The Biological Detection Kit. Proc. from the 6th International Symposium on Protection Against Chemical and Biological Warfare Agents, Stockholm, Sweden, May 10-15

32. Trudil, D., Loomis, L., Pabon, R., Hasan, J.A.K., and Garg, S., Trudil C.L. (2000) Rapid ATP Method for the screening and identification of bacteria in food and water samples. Biodetection Conference.(2000) February, 2-6

33. Ueda, I., and Suzuki, A. (1998) Is there a specific for anesthetics? Contrary effects of alcohols acids on phase transition and bioluminescence of firefly luciferase. Biophys. J. 75:1052-1057.

34. Van Crombrugge J.,Waes G., Reybroeck W. (1989) The ATP-F test estimation of bacteriological quality of raw milk. Neth.Milk Dairy,43, 347-354

35. Velazquez, M., Chan, H., Kirumira, A., Feirtag, J. (1996) Quenching and enhancement effect on the ATP bioluminescence signal using different ATP extractants and sanitizers. 83rd Annual Meeting of JAMFES, Abst. No.73, p. 50.

36. Velazquez, M., Feirtag, J. (1997) Quenching and enhancement effect on the ATP extractants, cleansers and sanitizers on the detection of the ATP bioluminescence signal. J. Food Prot. 60:799-803.

A RAPID METHOD FOR DETECTING BACTERIA IN DRINKING WATER

JIYOUNG LEE and ROLF A. DEININGER
Department of Environmental & Industrial Health
The University of Michigan
Ann Arbor, Michigan 48109

Abstract

A rapid determination of the total bacterial count in drinking water is important to the operators of treatment plants and distribution systems. It will allow corrective measures in real-time, such as increasing the disinfectant dose or removing water that has high bacterial numbers. The present heterotrophic plate count (HPC) analysis takes seven days and is not useful for operational intervention. The purpose of this study was to determine if a rapid adenosine triphosphate (ATP) assay would estimate the total number of bacteria in minutes. For quality control purposes and also to test the accuracy of both the ATP and HPC test, direct enumeration of the bacteria in a water sample was done using two epifluorescence methods. One was acridine orange direct count (AODC) method, which allows enumeration of both viable and nonviable bacteria. The other was direct viable count (DVC) method, which enumerates viable bacteria. Water samples originated from local, national, and international locations. The sample selection criteria were based on proximity to the laboratory, cooperating water utilities, and the travel of the authors. The results of the study show that the rapid ATP assay is highly correlated with the conventional plate count method and the DVC method, and estimates the bacterial quality of drinking water in minutes.

1. Introduction

The current microbiological standards focus on a single group of indicator organisms for the bacteriological safety of drinking water (6). Although the current standards of water quality have eliminated massive outbreaks of waterborne disease, a question has been raised about the adequacy of the standards of drinking water quality to prevent water-borne illnesses (16). Cases of gastrointestinal illnesses were reported among individuals drinking tap water that had met microbiological and physico-chemical water quality criteria (16). The determination of the total number of heterotrophic bacteria has been known as a better indicator of water quality than the coliform test because many opportunistic pathogens are not in the coliform group (7) and a high HPC has shown to interfere with the determination of the coliforms (10,11).

The present HPC method using R2A agar is known as the most sensitive test for enumerating bacteria from treated water (17). The disadvantage of the test is that it takes seven days to complete. Unfortunately, when the results are known, the water that was tested has long been consumed. A test is needed that determines the total bacterial populations in a very short time so that corrective actions can be taken before the water is consumed. Such a test is useful in the operation of a water treatment plant, the monitoring of the distribution system, the installation of new pipes, and the repair of old pipes. An additional application includes a potential investigation of an outbreak of a suspected waterborne illness to pinpoint the locations with high bacterial counts. If the information is available on a timely basis, control measures would be possible before the water is consumed. The control measures are increasing the disinfectant dose at the water treatment plant and removing water with a high bacterial level from the distribution system by opening some fire hydrants.

The ATP bioluminescence assay can serve this purpose. It allows an estimation of bacterial populations within minutes and it can be applied on site also. The estimation of the bacterial count based on the ATP of the water is not new. What is new is that our method is over 100 times more sensitive, requires one-hundredth of the sample volume,

P. Stopa and Z. Orahovec (eds.), Technology for Combating WMD Terrorism, 75-80.

and is over 10 times faster than the ATP method mentioned in Standard Methods (1995; 9211C.1) (1). Standard Methods indicate that the test requires one hour, one liter of water, and has a sensitivity of 100,000 cells/ml.

There have been studies in the literature that described failures and poor correlation between ATP and HPC (9,19). Such results were probably due to improper separation of the bacterial ATP from the nonbacterial ATP. On the other hand, a good correlation between ATP and HPC was reported when studying biofilms (20).

The purpose of this study was to determine if a rapid ATP assay could estimate the bacterial populations in a practical and timely manner. For quality control purposes and to test the accuracy of both the ATP and HPC test, a direct enumeration of the bacteria in a water sample was done using two epifluorescence methods. One was AODC method to enumerate the total number of bacteria, which include both the viable cells and the nonviable cells. The other was DVC method, which selectively enumerate viable bacteria.

2. Materials and Methods

2.1 WATER SAMPLES

Water samples were obtained from drinking water fountains or distribution system sampling locations in the United States and abroad. The samples in the United States were taken from locations in Michigan, Ohio, Illinois, Washington, D. C., California, Colorado, Florida, Georgia, Oregon, Washington, Kentucky, Tennessee, Maryland, Texas and New York. Some of the samples were obtained from airports (Illinois, California, Oregon, Washington, Kentucky, Tennessee, Maryland, Texas, New York), and some were obtained from cooperating utilities (Michigan, Ohio, California, Colorado, Georgia, Florida). A number of samples were obtained from Hungary, Germany, Switzerland, Netherlands, Austria, United Kingdom, France, Ukraine, Lithuania, Japan, Korea, Egypt, Saudi Arabia, Argentina, Peru, Brazil, Panama, and Australia.

2.2 FILTRATION OF WATER SAMPLES

A Filtravette™, which is a combination of a filter and a cuvette with pore size of 0.45 μm, was placed into a Swinex filter holder (13mm, Millipore Corporation, Bedford, MA). A sterile syringe was used for drawing the water samples. The testing volumes of the water were between 0.1 and 10 ml, based on the expected number of bacteria in the sample. The filter holder was screwed onto the syringe and the water sample was pushed through the filter. The Filtravette™ was taken out from the filter holder and placed onto a sterile blotting paper. The remaining water inside the filtravette was removed with a specially converted 3 ml syringe by applying gentle air pressure.

2.3 ATP BIOLUMINESCENCE

A somatic cell releasing agent (New Horizons Diagnostic Corporation [NHD], Columbia, MD) was used according to the manufacturer's instructions to lyse all non-bacterial cells and release ATP. Air pressure was used to remove the non-bacterial ATP through the filter. At this stage, the Filtravette™ retains bacteria on its surface, and the bacterial ATP remained within the bacterial cell membranes through this step of the procedure (5,18). This filtration method after special pretreatment was shown to be the most suitable for removing non-bacterial ATP (15). The Filtravette™ was inserted into the microluminometer (Model 3550, NHD, Columbia, MD) and the bacterial cell releasing agent was then added according to the manufacturer's instructions to lyse the bacterial cells retained on the surface of the filter. The released bacterial ATP was mixed with 50 μl of luciferin-luciferase (NHD, Columbia, MD) and the drawer of microluminometer was closed. The light emission was recorded after 10 second integration of the light impulses and the unit was relative light unit (RLU). The result was expressed as RLU/ml by dividing the RLU values by the filtered water volume. The detection limit and sensitivity of the luminometer was tested with serially diluted ATP solution (NHD, Columbia, MD). Distilled deionized water was used for the dilution and the pH was

7.8. The activity of the luciferin-luciferase was checked by using an ATP standard (NHD, Columbia, MD). The RLUs are proportional to the amount of ATP, and the amount of ATP is proportional to the number of viable bacteria (12).

2.4 BACTERIAL ENUMERATION: *AODC, DVC, HPC*

The total (nonviable and viable) bacterial cells were determined from formaldehyde fixed (2%, v/v, final concentration) samples with the AODC method (Hobbie *et al.*, 1977). The bacterial cells were stained with acridine orange (0.01%, w/v, Fluka, Switzerland) after filtration onto a 0.2 μm-pore-size black polycarbonate membrane filters (Poretics, Livermore, CA). Cells were enumerated at a magnification of x1000 with an Olympus Provis epifluorescence microscope (Olympus Optical Co., Japan) equipped with a mercury arc lamp and a 460 - 490 nm excitation filter. The number of bacteria was counted in 10 microscopic fields using three subsamples and was then averaged. The number of bacteria per milliliter of sample was calculated using the equation in Standard Methods (1).

The viable cells were counted by the DVC method (3) with some modifications. The samples were incubated with yeast extract (0.005%, w/v, Difco, Detroit, MI) and nalidixic acid (10mg/L, Sigma, St. Louis, MO) without dilution for 24 hours at 20°C. The modifications were using a lower concentration of yeast extract and no dilution. After incubation, the fixation, counting, and calculation of elongated bacteria were done following the AODC method.

The HPC was determined for each water sample in triplicate using R2A medium (Difco, Detroit, Michigan). The bacterial colonies were counted after an incubation period of 7 days at 28°C (1).

3. Results and Discussion

The detection limit of ATP was determined with high accuracy (r=0.999). It showed that the microluminometer was able to determine ATP down to 0.2 picogram. It is known that the average ATP content in one bacterial cell is about 10^{-15} g (1 femtogram) (4). Thus 0.2 picogram corresponds to 200 bacterial cells, which means the sensitivity of the ATP assay in terms of bacterial cell numbers.

About 120 water samples were analyzed with ATP bioluminescence method, HPC, DVC and AODC method, each in triplicate. The data show that there are high correlation coefficients between ATP, HPC, DVC, and AODC, and among those, the highest correlation coefficients were found between ATP and HPC (0.84), and ATP and DVC (0.80) (Table 1).

TABLE 1.
CORRELATION COEFFICIENTS BETWEEN ATP BIOLUMINESCNECE AND OTHER THREE PARAMETERS FOR DETERMINING BACTERIAL LEVEL IN DRINKING WATER

	ATP	HPC	DVC	AODC
ATP	1	-	-	-
HPC	0.84	1		-
DVC	0.80	0.75	1	-
AODC	0.68	0.66	0.73	1

The relationship between ATP and HPC is shown in Figure 1. In a number of cases where a cooperating utility sent water samples, the authors sent the expected HPC results by e-mail or fax on the day the water samples were received and analyzed. After the incubation period, the expected HPC and the measured HPC were compared. The expected HPC was calculated with the regression equation based upon the average ATP (RLU/ml). For instance, if the average value is 10 RLU/ml, the estimated HPC is 47 CFU/ml with a confidence interval (CI) between 32 (lower bound) and 71 (upper bound). In all but a few cases, the prediction and the measured HPC by the utility agreed. The regression equation is log HPC = 0.35 + 1.470 * log ATP (n=114). A factor that may affect the linear relationship between ATP and HPC (CFU) is the presence of injured bacteria which can not grow on agar plates.

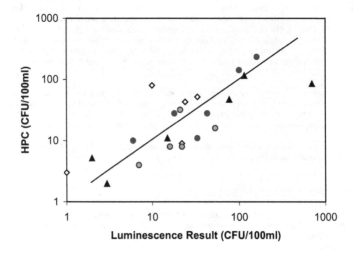

Figure 1. Correlation between Heterotrophic Plate Counts and Luminescence.

In treated drinking water, bacteria are exposed to disinfectants and a large proportion of the bacterial population became injured (14,2). Injured bacteria are viable but are not able to form colonies on agar plates (13); however, the injury does not directly affect the presence of intracellular ATP (13).

Theoretically, when ATP is zero with no detectable light emission, the direct viable count should also be zero. The intercept of the regression equation of 2.287 (log scale), meaning 193 cells, indicates the detection limit of the procedure; it cannot be detected when the cell number is less than 193. The relation between the HPC and the direct viable count is shown in Figure 1. The fact that the regression line does not go through the origin indicates that even when the HPC is zero, there are viable bacteria present. Thus, the estimate is that there are $10^{2.43}$ per ml, or about 270 bacterial cells which are viable, but not culturable on the R2A agar. The regression equation implies that when no detectable ATP occurs in the sample, a relatively large number of nonviable and viable cells are present and viewable using the acridine orange direct count method.

4. Conclusions

This is the first study where a miniaturized ATP bioluminescence method has been validated against the conventional plate method, the direct viable count, and the acridine orange direct count for the determination of heterotrophic bacteria in the drinking water samples. The ATP assay was found to be rapid and sensitive. The procedure is simple and can be done on-site with a portable power supply. The volume of water for filtration is small (0.1 - 10 ml). It could be applied in many other fields, such as the bottled water industry, the food industry, and wastewater treatment plants. Other applications include testing the water for military troops when they are deployed to an area where the potability of available water is in question. It is also applicable to examine ultrapurified water used in the pharmaceutical industry and the electronics industry. Due to its small size and its high sensitivity, it can also be used as a tool to monitor the biological safety of water in a spacecraft.

5. Acknowledgments

The authors are grateful for the participation of the water utilities that were involved in this study. In addition, the help provided by Dr. Martin Philbert and Dr. Peter G. Meier at the University of Michigan is sincerely appreciated. This project was partially funded by the Drinking Water Research Division, US EPA, Cincinnati, Ohio.

6. References

1. APHA-AWWA-WEF.1995. Standard Methods for the Examination of Water and Wastewater. 19th ed. American Public Health Association, Inc., Washington, D. C.

2. CAMPER, A. K. and MCFETERS, G. A. 1979. Chlorine injury and the enumeration of waterborne coliform bacteria. Appl. Environ. Microbiol. *37*, 633-641.

3. COALLIER, J., PREVOST, M., and ROMPRE, A. 1994. The optimization and application of two direct viable count methods for bacteria in distributed drinking water. Can. J. Microbiol. *40*, 830-836.

4. CROMBRUGGE, J. and WAES, G. 1991. ATP method. In *Methods for assessing the bacteriological quality of raw milk from the farm* (W. Heeschen, ed.) pp.53-60. International Dairy Federation, Brussels, Belgium.

5. CUTTER, C. N., DORSA, W. J., and SIRAGUSA, G. R. 1996. A rapid microbial ATP bioluminescence assay for meat carcasses. Dairy, Food and Environmental Sanitation. *16*, 726-736.

6. FEDERAL REGISTER. 1996. Vol. 61, No. 94. 24372.

7. GELDREICH, E. E., NASH, H. D., REASONER, D. J., TAYLOR, R. H. 1972. The necessity of controlling bacterial populations in potable waters; community water supply. J. Am. Water Works Assoc. *64*, 596-602.

8. HOBBIE, J. E., DALEY, R. J., and JASPER, S. 1977. Use of Nuclepore filters for counting bacteria by fluorescence microscopy. Appl. Environ. Microbiol. *33*, 1225-1228.

9. HOLT, D. M. and DELANOUE, A. 1997. ATP-can it measure bacteria in distribution systems. Proc. Water Quality Technology Conference. AWWA. Denver, Colorado.

10. LECHEVALLIER, M. W. and MCFETERS, G. A. 1985. Enumeration of injured coliforms in drinking water. J. Am. Water Works Assoc. *77*, 81-87.

11. LECHEVALLIER, M. W. and MCFETERS, G. A. 1985. Interactions between heterotrophic plate count bacteria and coliform organisms. Appl. Environ. Microbiol. *49*, 1338- 1341.

12. LECHEVALLIER, M. W., SHAW, N. E., KAPLAN, L. A., and BOTT, T. L. 1993. Development of a rapid assimilable organic carbon method for water. Appl. Environ. Microbiol. *59*, 1526-1531.

13. LITTLE, K. J. and LAROCCO, K. A. 1986. ATP screening method for presumptive detection of microbiologically contaminated carbonated beverages. Jour. Food Sci. *51*,474-476.

14. MCFETERS, G. A., KIPPIN, J. S., and LECHEVALLIER, M. W. 1986. Injured coliforms in drinking water. Appl. Environ. Microbiol. *51*, 1-5.

15. OLSEN, P. 1991. Rapid food microbiology: Application of bioluminescence in the dairy and food industry- a review. In *Physical methods for microorganism detection*. (H. Wilfred, ed.) pp. 64-80, CRC Press, Boca Raton, Florida.

16. PAYMENT, P., RICHARDSON, L., SIEMIATYCKI, J., DEWAR, R., EDWARDES, M. and FRANCO, E. 1991. A randomized trial to evaluate the risk of gastrointestinal disease due to consumption of drinking water meeting current microbiological standards. 1991. Am. J. Public Health. *81*, 703-708.

17. REASONER, D. J. and GELDREICH, E. E. 1985. A new medium for the enumeration and subculture of bacteria from potable water. Appl. Environ. Microbiol. *49*, 1-7.

18. SIRAGUSA, G. R., CUTTER, C. N., DORSA, W. J., and KOOHMARAIE, M. 1995. Use of a rapid microbial ATP bioluminescence assay to detect contamination on beef and pork carcasses. J. Food Protection. *58*, 770-775.

19. SMITH, E. S. and COLBOURNE, J. S. 1992. ATP rapid microbiology as a process control for slow sand filtration management. Proc. Water Quality Technology Conference. AWWA. Toronto, Ontario.

20. VAN DER KOOIJ, D., VEENENDAAL, H. R., BAARS-LORIST, C., VAN DER KLIFT, D. W. and DROST, Y. C. 1995. Biofilm formation on surfaces of glass and teflon exposed to treated water. Water Res. *29*, 1655-1662.

STRATEGIES FOR THE DEVELOPMENT OF NUCLEIC ACID PROBES FOR *WMD* AGENTS

Application of Genomics and Proteomics for Detection Assay Development for Biological Agents of Mass Destruction

VITO G. DEL VECCHIO
Institute of Molecular Biology and Medicine
University of Scranton, Scranton, Pa. 18510

1. Introduction

Distinguishing a particular microorganism from all others demands finding a target sequence that is unique for that organism, is conserved in all members of that species or strain, and has limited or no plasticity. Such targets can be a nucleic acid sequence, constitutive protein, or an antigenic epitope. The target can be probed with PCR-based, immunochemical, or mass spectroscopy assays. Genomics has played an important role in developing new generations of probes. The genomes of an organisms of which is the nucleic acid content has been sequenced and annotated is of invaluable aid in the identifications of targets. Post genomic disciplines such as proteomics and glycomics are and will usher in a new generation probes. The proteome can be defined as a set of proteins produced by an organism under a defined set of conditions. The presence of a protein is the ultimate proof that a gene is being expressed. The glycome represents the glycan groups or saccharide chains attached to proteins or lipids. Surface proteins (S-layer) are ideal targets. Since they are often found at the surface of a cell they are probed with antibodies without breaking the cell.

Genomic strategies that have been used in probe development assays include the uses of bioinformatics and *in silico* studies, fingerprints analysis, and suppressive subtractive hybridization (SSH). Proteomic approaches that are now emerging include global analysis in which the entire protein compliment is studied, comparative proteomics whereby the proteome of one species is compared to a closely-related species. Comparative proteomics has been used to investigate the secretome, which is the set of proteins secreted into a defined or semi synthetic medium. The secretome is sometimes referred to as the culture filtrate. The proteomes of an organism grown under different conditions can also yield valuable data for probe development. The protein content of a specific part of a cell such as the exosporium or S-layer is of great interest for these are often involved in attachment, internalization, and harming of host cells.

SSH is a PCR-base technique in which the genomic DNA of the organism of interest, referred to as the tester DNA, is hybridized with a reference DNA, termed the driver. The hybridized sequences that are common to the tester and the driver are then removed. The unhybridized sequences are tester-specific. These tester sequences are then cloned into *E. coli* for further investigation. SSH has been used to identify *Bacillus anthracis* chromosomal sequences that are not present in the closely-related *B. cereus* and *B. thuringiensis*. Figure 1 shows the relative locations of the *B. anthracis*-specific open reading frames (ORFs).

Various sequence found in these regions, cluster or isolated orphan ORFs have been observed by PCR amplification to be ideal targets for the identification of *B. anthracis*. Region I is composed of ORFs that had similarity with genes encoding enzymes and proteins involved in cell wall surface polysaccharides. Regions II and III contained bacteriophage-related sequences including integrases and recombinases.

The presence of a gene or ORF does not necessarily mean that a protein is produced. In fact only a small percentage of tRNAs are translated into proteins. Thus the presence of a protein is the ultimate proof that a gene or ORF is capable of being expressed. Proteomics has been the mains strategy for investigating gene expression on a global or comparative level. It is the premier post-genomic discipline.

P. Stopa and Z. Orahovec (eds.), Technology for Combating WMD Terrorism, 81-90.
© 2004 *Kluwer Academic Publishers. Printed in the Netherlands.*

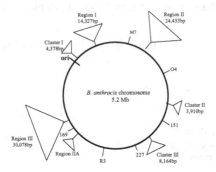

Figure 1. *B. anthracis* specific chromosomal components

2. Proteomics Methods

Standard proteomics studies utilize two-dimensional gel electrophoresis (2GE) in which the proteins are separated by isoelectric focusing (IEF) in the first dimension and by SDS-PAGE in the second dimension. 2GE is a powerful technique for resolving complex mixtures of proteins and permitting the simultaneous analysis of hundreds or thousands of proteins. IEF separates proteins on the basis of their isoelectric point (pI). The pI of a protein is the pH at which the protein has a zero net charge. When a protein mixture is applied to an IEF gel, each protein migrates until it reaches the pH that matches its pI. Proteins with different pIs align at different positions throughout the gel.

In the conventional method, proteins are separated in a pH gradient generated by applying an electric field to a gel containing a mixture of free carrier ampholytes. Carrier ampholytes are low molecular mass components (i.e., organic acids and bases) with both amino and carboxyl groups. When an electric field is applied to a solution of ampholytes, positively charged molecules migrate toward the cathode and those that are negatively charged move toward the anode generating a defined pH gradient. At its isoelectric point, an ampholyte cannot migrate because it has a zero net charge. When an IEF gel with a stable pH gradient is used, separation is achieved when each protein molecule migrates to the position of its isoelectric point and accumulates (focuses) there. Since the pH at each end of the gel is known, IEF is used to determine the isoelectric point of a particular protein.

After IEF, proteins are separated in the second dimension by electrophoresis in the presence of the detergent sodium dodecal sulfate (SDS). SDS binds to most proteins by hydrophobic interactions in amounts proportional to the molecular mass of the protein. The bound SDS contributes a large negative charge, rendering the intrinsic charge of the protein insignificant. Furthermore, when SDS is bound, most proteins assume a similar shape, resulting in a similar ratio of charge to mass. Thus, protein separation is based almost exclusively on the basis of mass, with smaller polypeptides migrating most rapidly.

After 2GE, the proteins are visualized by staining with SYPRO® Ruby, a ruthenium-based fluorescent stain shown to have several advantages over other commonly used protein stains with respect to sensitivity and linear dynamic range. The stained gels are imaged and the spots are picked for protein identification. The protein spots are subjected to protease (e.g. trypsin) digestion to yield smaller peptide fragments, the number and size of which are characteristic for a particular protein. The digested protein is then spotted onto a plate and identified by mass spectrometry (MS)

One commonly used instrument for most MS work is Matrix-Assisted Laser Desorption Ionization Time of Flight (MALDI-TOF) Mass Spectrometer. The mass analyzer resolves ions based on their mass/charge (m/z) ratio that is proportional to their velocity. This is dependent upon the time required for the ions to travel the length of the flight tube, i.e., the time between application of voltage to the plate and the registration of signal by the detector. The smaller the m/z value, the shorter is the flight time and the faster the ions reach the detector. The detector then converts the kinetic energy of the arriving particles into electrical signals. Resolution in mass spectrometry refers to the ability of the instrument to distinguish between ions of slightly different m/z values.

The mass spectrum generated from MALDI-TOF-MS is called a mass fingerprint which is characteristic for a particular protein. The identity of an unknown protein can be determined by comparing its peptide mass fingerprint with the theoretical spectrum generated by digestion of each of the proteins in a database by using a search engine such as Mascot. Protein identification from enzymatically-derived peptides depends on the frequency of specific cleavage sites within a protein. The cleavage sites yield a set of potential peptide masses that are unique to that sequence entry when compared to all other proteins entered in the database. A protein is identified when a significant number of the experimentally determined peptide molecular weights match the m/z values in the theoretical mass spectrum. Some of the databases commonly used for protein identification include NCBInr, SWISS-PROT, TrEMBL and OWL. Thus, a rapid and automated protein characterization is achieved. Alternatively and more advantageous in terms of speed and accuracy is matching the MS spectra with translations of the nucleotide sequence from an annotated genome of the same organism. This enables investigator to match expressed proteins to their corresponding open reading frames (ORFs) (Figure 2). For instance, the availability of a completely sequenced and annotated *B. melitensis* genome has paved the way for a highly comprehensive and rapid analysis of its proteome [1].

Figure2. Method of protein identification using the Mascot search engine

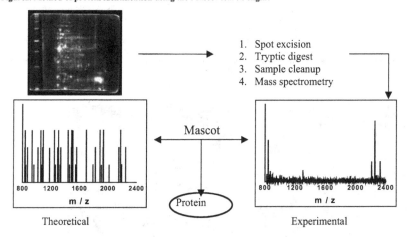

The results of proteome analysis indicate which genes are expressed under a given set of conditions, how protein products are modified, and how they might interact. Unlike the genome, the proteome is not static. It changes with the state of development, under conditions of environmental stress, and during disease states of a tissue. There are many more proteins in a proteome (mainly due to posttranslational modifications) than genes in a genome.

3. Bridging the gap Between Genotype and Phenotype

It is now possible to bridge the gap between genotype and phenotype. Using proteomics, we are not only able to determine what gene or ORF is responsible for a particular virulence property of a pathogen but also whether or not its product or protein is being expressed. Signature proteins can be applied to immunochemical assays to determine if a particular gene is being expressed. The gene that codes for that particular protein can be identified by various nucleic acid-based amplification assays. One must constantly keep in mind that the presence of a gene or its mRNA does not translate into the expression of its protein. Thus both nucleic acid- and protein-based assay must be used to determine if a particular virulence factor is present in a particular biological weapon of mass destruction (BWMD). This is extremely important in determining if a foreign virulence gene has been engineered into a BWMD.

3.1 *B. anthracis*

Knowledge of proteins that are induced during infection and those that contribute to pathogenicity would aid in the design of safe, efficient vaccines against *B. anthracis*, and could lead to the discovery of a new generation of effective anti-anthrax drugs. Such biomarkers could also be of immense aid in specific probe assays for *B. anthracis*. Using current proteomics technology, the proteins secreted into the growth medium in large quantities defined as "secretome" of the virulent *B. anthracis* have been analyzed under conditions that simulate an *in vivo* host cell. These conditions include the use of R-medium [2] that has been demonstrated to induce anthrax toxins and are totally synthetic so it facilitates the direct isolation of extracellular *B. anthracis* proteins. Secreted proteins have been isolated from the synthetic medium [3, 4]. Preliminary results indicate that twenty-seven proteins were detected in the simulated secretome. The well-characterized toxins, lethal factor, edema factor, the protective antigens as well as the known surface proteins were identified. In addition, a chromosomal *B. anthracis* enolase was identified which corresponded to an immuno-reactive protein in *Streptococcus mutants*. Overall, these findings could lead to the development of a new immuno-detection assay for *B. anthracis* and in the identification of stress factors and virulence-associated secreted proteins [5]. However, more investigations are warranted. Figure 3 shows the proteomes generated in induced and uninduced conditions.

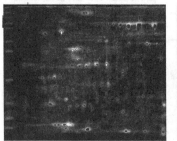 

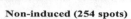

Non-induced (254 spots)   **Induced (322 spots)**

Figure 3. Overview of *Bacillus anthracis* RA3 secretomes (pH 4-7)

The secretomes of fully virulent, strains containing pXO1 but cured of pXO2; strains with pXO2 but no pXO1, and fully cured with no plasmids has been investigated to understand the pathogen response to simulated host conditions. Figure 4 is a representation of part of the secretome of the various cured and uncured strains.

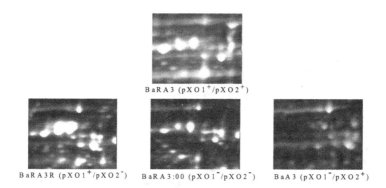

BaRA3 (pXO1⁺/pXO2⁺)

BaRA3R (pXO1⁺/pXO2⁻)   BaRA3:00 (pXO1⁻/pXO2⁻)   BaA3 (pXO1⁻/pXO2⁺)

Figure 4. Induced Secretomes of Virulent and Avirulent *Bacillus anthracis* strains

3.2 BRUCELLOSIS

Brucellosis is a major zoonotic disease that causes abortion and sterility in wild and domesticated animals and Malta fever in humans. Although the spread of the disease is controlled in developed countries by livestock testing, vaccination, and slaughter programs, brucellosis continues to be a major problem in the Mediterranean region and parts of Asia, Africa and Latin America where it causes severe economic losses [6]. *B. melitensis, B. abortus, B. suis* are the most common causal agents of brucellosis in humans. The World Health Organization considers *B. melitensis* as the most important zoonotic agent. It is extremely infectious (1-10 cfu per person) partly due to its highly aerosolic nature [7]. Because of its ability to cause a debilitating disease in humans, it is considered a potent biological warfare agent. The *Brucella* genus is composed of six currently recognized species: *B. melitensis, B. abortus, B. suis, B. ovis, B. canis,* and *B. neotomae* [6, 8]. Each species show preference in their host specificity to other animals. Recently brucellosis has been described in a variety of marine mammal throughout the world. Brucella are gram-negative, nonmotile, non-spore forming, coccobacilli [6]. The genome of one of its species, *B. melitensis* strain 16M, has recently been sequenced, closed, and annotated [1]. The genome is composed of 3,294,931 bp distributed over two circular chromosomes of 2,117,144 bp and 1,177,787 bp and predicted to encode for 3198 open reading frames (ORFs).

The global proteome of *B. melitensis* grown under laboratory conditions has been examined and is still a work in progress [1, 4, 9]. Figure 5 is a representation of the ORF expressed in laboratory grown cultures of *B. melitensis.*

Figure 5. Linear representation of what proteins are expressed under laboratory conditions by *B. melitensis.*

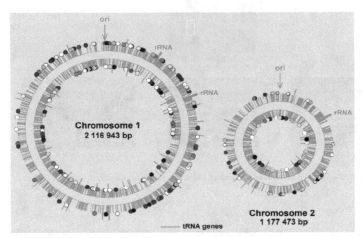

Figure 6 is an illustration showing the various classes of proteins represented by the different colored spheres.

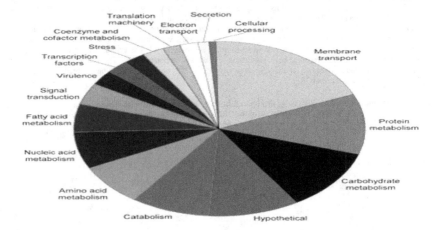

Secreted proteins often play an important role in the pathogenicity of microorganisms. It is therefore advantageous to identify which proteins are exported by a pathogenic bacterium and to acquire knowledge concerning how those proteins are exported from the cell. By genomic analysis, *Brucella* species are known to contain genes which share homology to those of known type I, III, IV, and V secretion systems. Brucella contains a conserved operon, *virB*, which has homology to type IV secretion system genes. In addition, 400-600 proteins of the *B. melitensis*, *B. suis*, and *B. abortus* theoretical proteomes are predicted to contain signal sequences which direct them to the periplasmic space, the outer membrane, or for export from the cell. While genetic analysis and computer modeling can provide insight to the potential of an organism, empirical evidence is necessary to confirm *in silico* hypothesis.

Such empirical evidence can be obtained by proteomic studies of the Brucella secretomes. Preliminary investigations have indicated that using a semi synthetic medium, which contains a minimum of complex biological nutrients, such studies can be achieved. Figure 7 is a 2DG of the secretome of various mutants of the virulence operon of *B. abortus*.

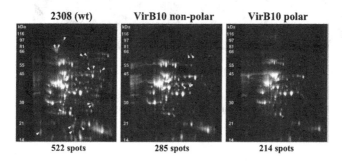

B. abortus secretome at pH 4 to 7
(major differences)

2308 (wt)   VirB10 non-polar   VirB10 polar

522 spots   285 spots   214 spots

▶ Protein spots missing in wt or mutant strains

▶ Differentially expressed proteins

Figure 7. Secretome of wild type and virulence operon mutants.

Membrane proteins are of special interest in the pathogenic process. The surface proteins are often the first to interact with the host cells. Many of the offensive mechanisms are found on the surface layer of a microorganism. Epitope biomarkers are easy to detect since the pathogen does not have to be lysed to probe. Surface proteins are often ideal vaccine targets. Figure 8 is a 2DG of total cellular proteins as compared to those found in the membrane of *B. melitensis*.

Once proteomics investigations have been identified potential biomarkers, probes can be developed to target specific nucleic acid and protein. Figure 9 is an illustration of this strategy

2DE of *in vitro* cultured *Brucella melitensis* 16M

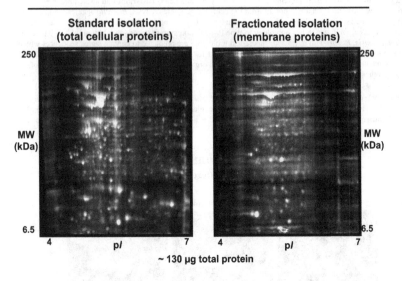

**Standard isolation
(total cellular proteins)**

**Fractionated isolation
(membrane proteins)**

Figure 8. Standard and membrane proteins of *B. melitensis.*

4. Conclusions

Advances in molecular biology, nucleic acid analysis, bio-analytical chemistry, mass spectroscopy, and bioinformatics are opening new insights into the analysis of bacteria. Current trends suggest that the old paradigm of searching for sequences that code for surface specific antigen markers or other structural components is no longer. There are a host of other "hooks" that one can use to design reagents for the detection and analysis of biological WMD agents. We are just learning what they are and how to get to them.

Figure 9. Proteomic biomarker discovery strategy.

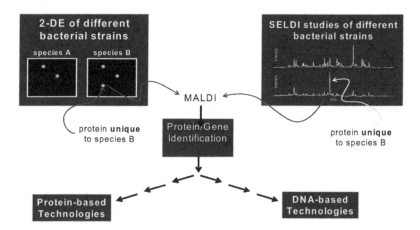

90

5. References

1. DelVecchio, V. G., V. Kapatral, R. J. Redkar, G. Patra, C. Mujer, T. Los, N. Ivanova, I. Anderson, A. Bhattacharyya, A. Lykidis, G. Reznik, L. Jablonski, N. Larsen, M. D'Souza, A. Bernal, M. Mazur, E. Goltsman, E. Selkov, P. H. Elzer, S. Hagius, D. O'Callaghan, J. J. Letesson, R. Haselkorn, N. Kyrpides, and R. Overbeek.(2002) The genome sequence of the facultative intracellular pathogen *Brucella melitensis*.Proc. Natl. Acad. Sci. USA. 99: 443-448.

2. Ristroph, J. D and B. E. Ivins. (1983) Elaboration of *Bacillus anthracis* antigens in a new defined culture medium," Infect. Immun. 39: 483-486.

3. Radie-Kolpin, M., R. C. Essenberg, J. H. Wyckoff.(1996) Identification and comparison of macrophage-induced proteins and proteins induced under various stress conditions in *Brucella abortus*. Infect. Immun. 64:5274-5283.

4. Wagner, M. A., M. Eschenbrenner, T. A. Horn, et al., (2002) Global Analysis of the *Brucella melitensis* proteome: identification of the proteins expressed in laboratory-grown culture," Proteomics 2:1047-1060.

5. Patra, G., Williams, L.E., Dwyer, K.G. and DelVecchio, V.G. (2003) *Bacillus anthracis* secretome. *In* Applications of Genomics and Proteomics for Analysis of Bacterial Biological Warfare Agents. Eds V.G. DelVecchio and V. Krcmery. NATO ASI Series.

6. Corbel, M. J., W. J. Brinley-Morgan, in Bergey's Manual of Systematic Bacteriology,   Eds. N. R. Krieg and J. G. Holt Williams and Wilkins (1984) 377-388.

7. Miller, C. D., J. R. Songer, and J. F. Sullivan (1987): A twenty five year review of laboratory-acquired human infections at the National Animal Disease Center," Am. Ind. Hyg. Assoc. J. 48 271-275.

8. Verger, J. M., F. Grimont, P. A. Grimont, and M. Grayon. (1987) Taxonomy of the genus *Brucella*. Ann Inst. Pasteur Microbiol. 138: 235-238.

9. Mujer, C.V., M.A. Wagner, M. Eschenbrenner, T.A. Horn, J.A. Kraycer, R. Redkar and V.G. DelVecchio. (2002) Global analysis of *Brucella melitensis* Proteomes: its potential use in vaccine development, identification of virulence proteins and establishing evolutionary relatedness. *In*: "The Domestic Animal/Wildlife Interface: Issues for Disease Control, Conservation, Sustainable Food Production, and Emerging Diseases." E. Paul J. Gibbs and Bob H. Bokma (Eds.). Ann. N.Y. Acad. Sci. 969:97-101.

CHEMICAL WARFARE AGENT SAMPLING AND DETECTION

PETER J. STOPA
Edgewood Chemical Biological Center
5183 Blackhawk Road
Aberdeen Proving Ground, MD 21010-5424

1. Introduction

First responders require appropriate equipment to sample and detect the presence of chemical or biological (CB) agents in the field. This paper will discuss some of the equipment that is currently available for the sampling and detection of chemical warfare agents (CWA) in the field.

2. Chemical Agent Detection – A Synopsis

Chemical agents may exist in several different physical states that include solid, liquid, aerosol, or vapor. One needs the appropriate sampling and detection equipment for each phase. Examples of this include scoops or spatulas for the solid phase; syringes or other types of liquid handling equipment for the liquid phase; and adsorbent for the aerosol or vapor phases. Several different manufacturers have assembled kits that may be used to sample for chemical agents in their various phases (Figures 1 and 2).

Another important factor is that chemical agents, under the appropriate temperature and pressure, do have an intrinsic vapor pressure. Thus different types of detectors may be used to determine the approximate places to sample for these materials.

As previously stated, chemical agents may exist in various physical states. One will also require not only different sampling schemes, but also different detection schemes specific for the state of the agent. For example, suspect liquids will require a means to determine if the liquid is a chemical agent or not. Fortunately, there are simple paper test strips that are available and have been in us for some time to determine this. These papers generally work on a density displacement method whereby the liquid sample will displace a dye that is trapped within the pores of the paper. In the US and elsewhere, a general indicator paper is called M-9 paper whereby a dye is displaced from the paper. This paper does not possess a means to differentiate among the class of agents and may also be prone to false positive reactions from liquids that possess a similar density. M-8 paper, on the hand, does offer a degree of differentiation among the G, V, and H agents. This paper is also prone to false positive reactions from liquids that have similar densities to the agents themselves (Figure 3).

There a variety of detectors for the vapor form of chemical warfare agents. Perhaps the simplest of them all is the M-256 detection kit, which utilizes a variety of wet chemistries, each specific for a class of agent. These different tests are grouped together on a detector card and one simply breaks a series of vials to detect the presence of the agent vapor.

The next step up in complexity is tube-based detectors that utilize agent specific chemistries that are immobilized on absorbent tubes. One simply draws an air sample across through the tube and awaits a color change to determine the presence of a particular class of agent. For example, there are specific tubes for organophosphates, thio-ethers, cyanide, chlorine, and many other classes of materials. One can simply gang together several of these tubes and have a detection system that is capable of detecting several classes of agents simultaneously.

P. Stopa and Z. Orahovec (eds.), Technology for Combating WMD Terrorism, 91-94.
© 2004 *Kluwer Academic Publishers. Printed in the Netherlands.*

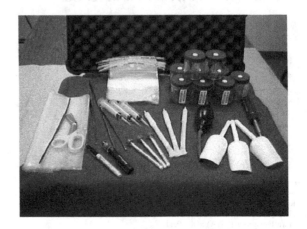

Figure 1. An example of a chemical agent sampling kit.

Figure 2. A CWA sampler for vapors.

The next step up in complexity is tube-based detectors that utilize agent specific chemistries that are immobilized on absorbent tubes. One simply draws an air sample across through the tube and awaits a color change to determine the presence of a particular class of agent. For example, there are specific tubes for organophosphates, thio-ethers, cyanide, chlorine, and many other classes of materials. One can simply gang together several of these tubes and have a detection system that is capable of detecting several classes of agents simultaneously.

Figure 3. M-8 Paper for liquid CWA detection.

There are a variety of instrumented techniques that utilize several different mechanisms to detect chemical agent vapor. Perhaps the most widely used is ion mobility spectrometry, which is similar in principle to mass spectroscopy. A vapor is drawn into an ionizing source whereby the components within the vapor are broken down into several different low-resolution mass ions. These ions then traverse a drift tube and are then separated by molecular weight. These specific ions are then detected on a detector tube and results are then tabulated. There are several commercial versions of this technology.

Another approach uses surface acoustic wave technology and agent-specific coatings. In this approach, the vapor is drawn into a chamber that has a piezoelectric crystal in it. On the surface of this crystal are specific coatings for each of the different classes of agents. The crystal then measures a weight change in the coating. By analyzing the resultant pattern from the dissolution of the agent within the coating, one may be able to determine the type of agent that is present.

Figure 4. Example of a passive standoff detector.

Field portable mass spectrometers have also been utilized for chemical agent detection. This represents the most sophisticated (and expensive) means to detect chemical agents; however, this may not be the most sensitive way to detect certain classes of agents.

All of the approaches described to now can be classified as "point" detectors. What this means is that one is detecting the agent at a point and thus needs to place the detector directly in the agent cloud. Another approach exists, however, whereby one may "stand off" from the cloud and detect its presence, hence the name "Standoff" detection. Two mechanisms are predominately used: active and passive approaches. In the active approach, one uses a spectral source (typically an infra-red laser) to scan an area. A signal then bounces back from the cloud to a detector system. With this system one can discern the direction and range to the cloud, similar to a radar system. This system is called "LIDAR" – Laser Direction and Ranging. One may also obtain physical or chemical information on the cloud, depending the types of laser(s) that are used in this system.

There is also a passive system standoff system that is widely utilized. A telescope is placed in front of a Michaelson interferometer and specific spectral regions are scanned. The presence of a signal, as compared to a background, denotes the presence of an agent. Typically temperatures are used as a means of detection. This approach is the most widely used standoff approach.

94

Perhaps one of the most important questions is how do the sensitivities of these detectors correspond with the particular LD$_{50}$, Permissible Exposure Limits (PEL), and concentrations that pose an Immediate Danger to Life and Health (IDLH). These results are summarized in Figure 5.

Performance of Various Chemical Agent Detectors

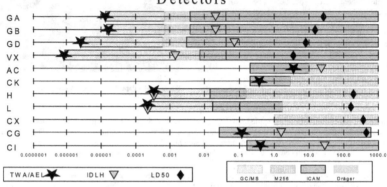

Figure 5. Performance of various chemical agent detectors with respect to detection limits. The detection limits are compared to the concentrations for the Lethal Dose (LD$_{50}$); Immediate Danger to Life and Health (IDLH); Time Weighted Average (TWA) or Airborne Exposure Limit (AEL).

The figure shows several things. Each of the detectors have different sensitivities for different things and that these sensitivities vary. Most of the detectors systems can detect at the LD$_{50}$ level, but then their performance varies when compared to either the IDLH or the TWA. The results also show that detectors based on relatively simple technologies can have good sensitivities for various agents.

3. Summary

In summary, there are a variety of technologies available for the detection of chemical warfare agents. One needs to keep in mind that these agents fall into a variety of classes and physical states. The performance of these detectors varies by agent, although some relatively inexpensive approaches can be effective. Therefore, the type of detector that one should pick should be dictated by the sensitivity that one needs and the operational scenario.

NEW GENERATIONS OF PROTECTIVE EQUIPMENT FOR RESCUE OPERATIONS FOLLOWING TERRORIST CHEMICAL, BIOLOGICAL, RADIOLOGICAL AND NUCLEAR ATTACKS

JIRI MATOUSEK

Masaryk University Brno, Faculty of Science,
Research Centre for Environmental Chemistry and Ecotoxicology
Kamenice 3, CZ-625 00 Brno, Czech Republic

Abstract

This paper presents new progressive generations of Czechoslovak/Czech equipment for personal protection that was originally developed and produced for both the armed forces and civil protection. This equipment is suitable for rescue operations during any emergency where toxic, contagious agents, and radionuclides; are released, either during wartime and peacetime, including chemical, biological, radiological and nuclear (CBRN) terrorist strikes. The complex system of these protective measures has been designated for first line rescue teams and simultaneously covers needs of protection for victims of all age groups

1. Introduction

Rescue operations, during both wartime and peacetime, where the release of toxic chemicals, bacteriological (biological) and toxin agents, and radionuclides need a broad spectrum of protective means for various risk levels as well as high professional, physical, psychological, and moral preparedness of responders. This type of equipment includes protective masks and various types of methods for protecting skin, which was originally developed for armed forces, civil protection, and fire brigades. It is understood that protection should be provided not only for rescue teams but also for the potential victims within the range of the plume of agent(s) released. Rescue teams choose the respective methods/equipment depending on the circumstances of a given event, i.e. the level of risk the event poses.

The Czech Republic/Czechoslovakia has long traditions in the research and development (R&D), and production of protective equipment. This tradition already started in the early 1920s, shortly after the formation of Czechoslovakia on the debris of the former Austro-Hungarian Monarchy, since it was surrounded by no friendly state when two neighbors were preparing offensive chemical assets. This work also continued after WW-II and during the Cold War, when this country (together with the then two German states) was located on the divide between two major military political alliances and could have been the potential target of weapons of mass destruction by both sides. That is why Czechoslovakia/Czech Republic has belonged to states with highly developed and continuously improved systems of protective equipment for the armed forces, civil protection, fire brigades and industry. This paper presents a discussion of new generations of military protective masks, protective masks for the civilian adult population, and protection equipment for children of all age groups.

Protective suits of all kinds are also employed. Besides classical protective suits (i.e. isolation types), single-use suits are also used. This group of progressive suits includes air-permeable protective suits and also filter-ventilated suits for heavy-duty operations under high-risk level. These suits also provide a means for operations under diminished oxygen concentration in ambient air.

P. Stopa and Z. Orahovec (eds.), Technology for Combating WMD Terrorism, 95-101.
© 2004 *Kluwer Academic Publishers. Printed in the Netherlands.*

2. Military and Civilian Protective Masks for Adults and Children

2.1 PROTECTIVE MASKS FOR THE ARMED FORCES

The tradition of manufacturing military protective masks in Czechoslovakia started in 1923 (while the first military mask introduced in 1919 was of French origin). The top pre-WW-II military protective mask was the Model 35. It had a face-piece made from high-quality rubber and a high-volume canister placed in a carrier bag. It had a very progressive design, especially for the then non-traditional position of visors, which assured the highest possible field of vision and used a leading filtered air-stream against the visors, thus preventing their fogging. This type, as was a series of pre-WW-II civilian masks, was produced in the new factory Fatra Napajedla, which belonged to the Bata boots concern in Zlín. This mask was also produced under license in Yugoslavia till the 1960s. Shortly after WW-II, the Czechoslovak Army used, in addition to this type, mainly standard Wehrmacht gas-masks. In the late 1940s/early 1950s, the production of new military masks was initiated in the enlarged factory Gumárny Zubri that was one of the pre-WW-II producer of civilian protective masks, starting with the Leyland-licensed model in 1936. This all-head face-piece type was introduced according to the Soviet-originated model SM-41M. It had also a large-volume canister in a carrier-bag and was connected by a hose with the face-piece and two exhaust valves. This design, together with the all-head concept, assured a very low coefficient of inward leakage.

In the early 1960s, research on new sorption materials was initiated that would enable wide utilization in various protection systems. The primary goal was to construct light and efficient filters against vapors and aerosols for a standard military protective mask of a new design as well as for some other (civilian) purposes. The main result of the R&D was a completely new type of military protective mask, the **M-10**. It had two face-piece filters (inspired by the US model M 17) introduced in the early 1970s. For the manufacture of rubber parts, a new pressurized-injection rubber-technological procedure was developed.

The combined filter material was the result of several R&D approaches. The best solution was eventually found. It was based on several layers of randomly laid, mixed polymer and other fibers. It was prepared by using a non-woven technology that pneumatically laid and then fixed tiny particles of specially prepared charcoal to the fibers.

This type of mask was considerably upgraded in the 1980s, designated as the **M-10M**. These upgrades included fitting the mask with a drinking device; new material for visors; additional visors for low temperatures; a protective hood; etc. Both types are still in use. The Czech Army also possesses some other special types of masks, such as the **SR-2**, for use on the heads of wounded persons; special mask **PRV-U** (with small round visors) for working with special optical devices; and the breathing device **PPS-500**.

In 1996, a new military protective mask, designated as the **OM-90**, was introduced [1,2]. This face-type protective mask, when furnished with a clamping system and the filter canister OF 90, complies with the best current world standards. Its face-piece (made from bromobutyl-rubber in black color – which, by the way, this the first Czech mask made in black) is extraordinarily resistant to the penetration of toxic chemicals. Design of a sealing gasket and low respiration resistance (pressure loss at air flow of 30 L/min: inhalation max 20 Pa, exhalation max 60 Pa) ensures long-time endurance of the mask with a filter canister made of high-resistant plastic. The design allows absolutely perfect orientation due to computer-optimized field of vision (general field of vision 71 %). The design enables (like at all previous masks) correction of dioptrical defects and compatibility with basic military optical devices including the night vision system "Clara". Hardened glass visors are very resistant to impact and scratching. The mask enables safe drinking in a contaminated area. A built-in diaphragm assures easy communication and audibility. Materials warrant their resistance to wear and damage and enable decontamination and simple maintenance. The face-piece is produced in 3 sizes, with an average weight of 500 g (approximately one half pound). The screw-on filter uses 40 x 1/7" thread and can be fitted on both the right and the left. The plastic filter canister (OF 90) is 110 x 80 mm, weighs 250 g, and exhibits a pressure loss at 30 L/min of a max 130 Pa. It has a very small particle penetration coefficient KP 0.0001 %. The OM kit contains also the single-use protective set JP-90 (see below).

2.2 CIVILIAN PROTECTIVE MASKS FOR ADULT POPULATION [2,3,4]

Czechoslovak production of civilian protective masks was started in the mid-1930s in three factories. As a result, four types of very cheap civilian masks were available on the eve of the WW-II. This corresponds to when the whole population was ordered by the Government (in 1938) to self-provide a protective means, vis-à-vis the Nazi threat of chemical war. This order was confirmed after German occupation in the Bohemia and Moravia Protectorates (1939). Fortunately, the Czech population survived WW-II with these protective masks without a necessity for their use. Civilian protective masks were developed in the post-WW-II era to provide for workplace safety and to protect against harmful chemicals in developing chemical and nuclear industries. Further development of WMD during the beginning of the Cold War and the East-West confrontation resulted in the effort to equip every citizen with protective equipment, in the first line with protective masks by government funding. This was the impetus for starting R&D and mass production of protective masks and equipment for all population groups. The ambitious task to equip all of the population with civilian masks was finally achieved in the 1960s. The first post-WW-II civilian protective mask for adult population was the simplest type (CM-3), stemming from the mid-1950s. It was produced in 4 sizes, covering the needs of the adult population (after 12 years of age). This mask, like all further types of civilian mask, is the face-type.

The considerably upgraded (nowadays basic type) is the **CM-4.** It uses some elements of the formerly introduced M-10, such as the curved visor that is made from more resistant material, and the voice-transmission membrane. It was produced since the 1970s is still the mostly used Czechoslovak/Czech civilian protective mask. From it, the upgraded **CM-4M** that is fitted with a drinking device was derived. An alternative type, the **CM-4K,** was also produced. Both stem from the 1980s.

The most up-to-date civilian protective masks (introduced in 1997) are equipped with one panoramic visor. They are the **CM-5** and **CM-5D**. Corresponding types that enable drinking without braking the mask seal are the **CM-5M** and the **CM-5DM**. The newest civilian protective mask, the **CM-6**, also has one panoramic visor and other construction and material improvements. It also has a very elegant overall design was sent into production recently (2002).

3. Protective Masks and Equipment for Children of all Age Groups [4,5]

Protective masks for children were already produced on the eve of WW-II. However, the complex solution of protecting all child age groups emerged as the response to challenges of the threats posed by the Cold War, as previously mentioned.

3.1 CHILD PROTECTIVE MASKS

Our child protective masks are generally designed for children between 3 and 12 years of age. The standard child protective mask **DM-1** (produced in three sizes: 0,1,2) is derived from the basic type of protective mask for adults (CM-3). In order not to excessively overburden the child's neck and spine, the filter canister is connected to the face-piece by means of a hose that is worn in a carrying bag. The smallest size (0) can be alternatively used even for the smaller children (18 months – 3 years). On the other hand, for the more developed children between 10 – 12 years another solution exists. It is the intermediary type **CM-3/3h**, i.e. the smallest size for adults with the canister filter in a carrying bag. A fully new type of child protective mask, the **DM-5,** has been developed recently. It is derived from the **CM-5** mask.

3.2 CHILD PROTECTIVE JACKETS

Protective jackets are generally designed for children between the ages of 1.5 – 3 years. We took into consideration the known problems that children have with wearing such strange things like protective masks. Therefore, we developed protective jackets that protect the head and the whole upper part of the child's body. The older type **DK-62** from the early 1960s is made from a rubber-coated textile fabric and is fitted with a panoramic visor from transparent plastic in the face part. This jacket uses a hand-operated fan to lead fresh air over a standard canister filter. It is connected with the under-jacket space by means of a hose.

The new type **DK-88** introduced in the late 1980s is made from light plastic material. A battery-operated ventilator delivers clean air through a standard canister filter that is mounted on the rear head part of the jacket. It is also fitted in the face piece with a large panoramic visor made from a resistant plastic. The exhaust overpressure valve is located under the visor. A light source for the ventilator uses by 6 x R-14 (or LR-14) batteries and is located in the breast pocket. The jacket is fitted with a drinking device.

3.3 CHILD PROTECTIVE BAGS [4,5]

Protective bags are generally designated for the smallest children, from newborns until 18 months of age. They are used for temporary protection of children before reaching shelters or evacuation places with filter-ventilation. The older type **DV-65** (from the mid-1960s) is manufactured from rubber-coated textile fabric on a metallic carrier frame. It possesses one large-area air-permeable diffusion filter on the sidewall that protects against toxic, radioactive, and biological agents. This is also how water vapor and exhaled carbon dioxide are expelled. There is one plastic visor on the upper part and manipulation gloves on the sidewall.

The recent type **DV-75** (introduced in the late 1970s) is considerably upgraded. It possesses two large-area air-permeable diffusion filters on both sidewalls. The manipulation gloves are also doubled. This type is fitted with a device to fill the suction bottle located inside. In the interior, there are two pockets for storing diapers. Accessories also include instructions for use of the storage bottle with cover, lid, and injection syringe; suction bottle with rubber-teat; carrier strap; inflatable pillow; and specially formed bed-linen which can be used as a child bed in the shelter. This is accomplished by stretching the linen over a supplied frame.

4. Filter Canisters for Protective Masks

Current filters for military and civilian protective masks are universal, considering that they can protect against all types of WMD. They can be used for OM-90 (standard filter **OF-90**) and for all types of civilian masks for adults and children (including the child jacket). The standard series of stored and used filter canisters is represented by **MOF-2**, **MOF-4**, **MOF-5** and **NBC-1**. The recently introduced filter **MOF-6M** protects also against volatile Toxic Industrial Chemicals (TICs). All masks are compatible with the internationally standardized (EN) series of filters against industrial chemicals (thread 40 x 1/7 "). The traditional Czech producer of these filters, since the 1930s, is the known Sigma Group Co in Olomouc – Lutín [6].

5. Skin Protection – Protective Suits

5.1 CLASSIC ISOLATING PROTECTIVE SUITS

The variety of tasks under contemporary combat, other wartime operations, and peacetime conditions; the ongoing implementation of Chemical Weapons Convention (CWC) and the Biological and Toxin Weapons Convention; and protection against the accidental release of hazardous materials, including those that might be caused by terrorist attacks, have contributed to better defined risks levels. This has resulted in a wider assortment of protective suits, the requirements of which result in a higher degree of physiological comfort, thus enabling long-lasting operations in an environment contaminated with toxic, biological and radiological materials, without compromising the integrity (seal) of the suit. This is reflected by new developmental trends, oriented to air-permeable protective suits for low to medium risk operations and to several types of heavy-duty, isolating protective suits, the foremost of which are filter-ventilated suits.

Based on the experience with the production of protective suits for the specialists in the inter-war period, the first generation of protective suits was introduced shortly after the founding of the Czechoslovak Chemical Corps in 1949. More appropriate models with better properties were developed in the 1960s. The classical Czechoslovak/Czech heavy isolating protective suits **OPCH-70** (Army)[7,8] and **SOO-CO** (Civil Protection)[8,9] are still in use. They have the same design, but differ in the material used. The combat variant is designed to protect against chemical warfare agents (CWA), while the civilian one was introduced later against a broader spectrum of chemicals, including toxic industrial chemicals (TICs). Both these suits consist of a protective overall (with hood), protective gloves, and protective boots and/or over-boots.

5.2 PROTECTIVE SUITS FOR SINGLE USE [7,10]

The oldest and simplest means for preventing CW contamination with persistent agents, at the time the most likely mode of contamination (aerial spray), was a large cape of impregnated and laminated paper that contained simple handles within, protecting against droplets (early 1950s). This item constituted the main component of the individual protective ensemble, which also included long gloves and long over-boots, made from a rubber-coated fabric.

The most sophisticated protective ensemble is represented by the protection set **JP-75A** (from the mid-1970s), which provides protection against droplets of CWA and radioactive dust, as well as short-term protection against incendiaries and a low intensity pulse of nuclear explosions. This set consists of a cape, gloves, and reinforced over-boots. It is manufactured from high-pressure polyethylene with self-extinguishing properties. This system has a minimum protection time of 20 min and is still in use, assuring minimum protection or additional protection to that of air-permeable suits.

5.3 AIR-PERMEABLE PROTECTIVE SUITS

The physiological burden of wearing a heavy isolation suit in a warm environment is significant affects the span of operating time. The blockage of the evaporation of vapors and heat causes an increase of the heart rate and body core temperature (measured as rectal temperature) that can be fatal when exceeding $38 - 38.5\ ^{\circ}$ C (and even lower with people who have a compromised thermoregulatory system). These physiological as well as ergonomic factors were the impetus behind the development of new suits for higher physiological comfort under the respective protective properties required for relevant risk levels. However, the time for operating in a contaminated environment must be comparable to that of the classic heavy isolation suit. These considerations plus the final goal to considerably extend operation times were the goals for our R&D in the two main new trends in protective suits for future uses, i.e. air-permeable suits and filter-ventilated suits.

We started the R&D on air-permeable (or adsorption) protective suits in the late 1960s, being inspired by our success in the manufacturing of the face-filters for the protective mask M-10 and similar technological approaches for the large-areas diffusion filters for the children bag DV-65. In both described cases, the technological solution was based on several layers of non-woven fabrics containing randomly laid polymeric fibers with pneumatically-fixed tiny particles of high-efficient charcoal. The garment for the air-permeable suit of our design consists of three layers (in the direction from the body surface):

1 - light air-permeable linen with good mechanical properties (protecting the second layer).
2 - main functional adsorption layer.
3 - upper cover layer with good mechanical properties and additional military/chemical resistance properties.

Because of the consideration of the air-permeable suit as the standard protective suit for every soldier, the upper cover was subjected to many additional requirement, such as water-repellence; impregnation against lipophilic agents (such as oil products and dirt); fire-retardance; camouflage prints; etc. It was also necessary to develop corresponding accessories, reflecting both material and operational requirements (i.e. protective gloves and over-boots). The result was the standard air-permeable over-suit **FOP-85** (introduced under original designation **POO** in the mid-1980s), consisting of a long blouse with cap, trousers, gloves, and over-boots (overall weight 3.5 kg, protection time against vapors and aerosols of CW-agents $6 - 24$ hrs, resistance against a light nuclear pulse 60 J/cm^2). As with any other protective suit, this ensemble underwent extensive physiological testing, showing a considerable decrease in the physiological burden as compared with the classic isolation suit. This suit is very comfortable to wear due to the wicking away of water vapors and heat. This enables medium to hard work for $3 - 6$ hrs in an ambient temperature of about 30° C [11,12].

From the basic suit, several variants have been derived [13]. They differ according to the customer's wishes, mainly in the construction, material, and color of the cover garment. In some cases, the suits are without the standard treatments against water, fire, dirt, and camouflage prints, enabling work in a hot climate. The Czech UNSCOM inspectors in Iraq successfully used one such suit, depicted as the POO-A. Its properties were highly appreciated by inspectors of some of the other nations that had the opportunity to use it or to conduct some informal testing of it. Another variant, designated as the **FOP-90,** is produced for Slovak Civil Protection. Yet another upgrade (designated as the **FOP-96)** was done for the Czech Army recently [11,13].

5.4 FILTER-VENTILATED PROTECTIVE SUITS FOR ARMED FORCES

The air-permeable suits, in spite of their high physiological comfort, cannot substitute for classical isolation suits, especially in cases of very high concentrations of agents, contamination with large drops of super-toxic lethal agents, or in cases of spills of such agents, clearly the most dangerous. Experience with use of heavy isolation suits has shown that the efforts to try to extend the stay in these suits, e.g. by various methods of cooling, have only limited effects. The physiological burden is mainly based on the impediment of the evaporation of water vapor and thermal energy of the human body. Our idea was to solve the problem through continuous delivery of fresh (filtrated) air into the under-suit space, creating also an overpressure. This simultaneously provides an elevated level of protection against possible leakage in normal sealing areas or against small amounts of piercing of the suit.

We have applied the best available technology to the protective garments with the highest resistance (tested for HD and GD in both static and dynamic conditions) for the construction of filter-ventilated suits. These technologies include special connection of the body of the suit with the hood, utilizing experience gained from divers and tight double-sealed lines at the extremities (gloves, boots) and head (protective mask). The battery-operated filter-ventilation unit (FVU), carried on the back (tested dynamically for CW, such as HD, GB, GD and TICs, such as CG, AC, CK, as well as for aerosols) delivers a regulated volume of fresh filtrated air (up to 300 L/min) over a distributor to the mask (providing for additional protection) and to the under-suit space be means of screwed hoses. 6 exhaust valves located on the suit body and extremities control the standing overpressure. The capacity of the battery is calculated for average operation (8 hrs). When the exchange of batteries occurs, nothing happens. The suit works like the classic isolation suit. The physiological comfort enables an incredible extension of the operating time, provided that the protective mask with drinking device (such as e.g. M-10M, OM-90 and the like) is used. We have extended very considerably the operation time using a urine-catchment device, consisting of a special collection device (only for the male population) that is connected to a hose and a small container located at the outer fibular part of the leg, enabling relief without breaking the seal.

Rigorous ergonomic testing under standardized conditions has shown very good results in the dramatic elongation of operational time. It sounds nearly incredible: we tested this suit for in medium to hard work, alternated with rest and sleeping, for 24 hrs. For the same work intensity without breaks, the wearer is able to work for 3 hrs. Overall weight was originally 11 kg. We have introduced this suit under the designation **OPCH-90** in the early 1990s [14]. The suit is fielded and now routinely used by the specialists of the Chemical Corps. The original filter-ventilated unit FVU-90 was later upgraded (FVU-95) to diminish its weight and give it a more flat form to enable entering narrow spaces. It is possible to note that this suit was used by the Czech inspection teams in the warm environment of the UNSCOM missions in Iraq and highly appreciated for its comfortable wearing under such conditions.

5.5 OTHER HEAVY-DUTY PROTECTIVE SUITS [14,15]

The basic filter-ventilated suit OPCH-90, used by military specialists (in brownish-green color), was adapted for other uses, including civil protection, fire brigades, and the chemical and nuclear industries. The suit, modified in the 1990s for civil protection and industrial uses, designated as the **OPCH-90CO,** is made from a different material that is resistant to oil products (hydrocarbons, organic solvents etc.). It is manufactured in bright (signal) yellow and utilizes a rounded-shape, very flat filter-ventilation unit. We envisage using this suit for a wide-variety of applications. A training variant has also been developed that uses a lighter material.

The OPCH-90CO was modified as **OPCH-90PO** for use by fire-brigades and rescue teams operating in an environment with an oxygen content below 17 %. It uses an external source of oxygen in a steel bottle instead of a FVU suitable for use in the most hazardous conditions, such as a corrosive environment or in an area contaminated by highly contagious biological agents. The oxygen source is carried under the suit to provide for extraordinary high assurance against inward leakage.

6. Conclusions

Due to the long Czechoslovak/Czech traditions and experience with its own systematic R&D, production and use, the Czech Republic possesses a wide and complete assortment of continuously upgraded personal protection equipment for the armed forces, civil protection, and industrial uses. This equipment is available for emergency situations, either under wartime conditions or peacetime releases of toxic, radiotoxic, and contagious materials. The first line of defense is the military protective mask as well as protective masks for adult civilians and equipment for children of all age groups. Various levels of skin protection are provided by several types of air-permeable protective suits or by filter-ventilated suits. Therea are derivatives of these suits that can provide assure long-term work under the most hazardous conditions. The domestic R&D has also led *inter alia* to the well-developed system of testing personal protection equipment, encompassing a sophisticated set of tests for technical, protective, physiological and ergonomic parameters, that are compatible with NATO and European standards (EN), and are performed in an accredited and internationally renowned laboratory [16].

7. References

1. Matousek J.: Military protective masks (in Czech). *Rescue Report* **5**, No 2, 12 (2002).

2. Gumárny Zubri: Prospect materials (1999-2002).

3. Matousek J.: Civilian protective masks for adult population (in Czech). *Rescue Report* **5**, No 3, 12 (2002)

4. Institute for Protection of Population: *Means of Individual Protection for Civil Population of the Czech Republic* (in Czech), IPP Lázne Bohdanec, 2001.

5. Matousek J.: Protective masks and means for the children population (in Czech). *Rescue Report* **5**, No 4, 12 (2002).

6. Sigma Group, Olomouc: Prospect materials (2001-2002).

7. Army of the Czech Republic: *Catalogue NBC Defence and Chemical Support Equipment.* The NBC Monitoring Centre, Prague 2001.

8. Matousek J.: Classic isolating suits (in Czech). *Rescue Report* **5**, No 5, 12 (2002).

9. see 4.

10. Matousek J.: Single-use protective suits and improvised means (in Czech). *Rescue Report* **5**, No 6, 13 (2002).

11. Matousek J., Obsel V.: Technologies for air-permeable protective suits. International Conference Community – Army – Technology – Environment, CATE-95, Brno 1995. *Proceedings*, Section 4, p 49.

12. *Janes's NBC Protection Equipment 1995-1996.* 8th Ed. Jane's Information Group, Coulsdon 1995.

13. BOIS-Filtry, Brno: Prospect materials.

14. Matousek J., Slabotinsky J.: Filter-ventilated non-permeable protective suits – unique solution assuring the long-term work in contaminated area. The 5th International Symposium on Protection against CBW Agents, Stockholm 1995. *Proceedings*, pp 171-2.

15. EcoProtect, Zlín: Prospect materials

16. Matousek J., Slabotinsky J., Bradka S.: System of testing means of personal protection. 6th International Symposium on Protection against CBW Agents. Stockholm 1998. *Proceedings*, vol.1, pp 123-7.

17. The author is a former Head of the Czechoslovak NBC-Defence R&D Establishment.

PERSONAL DECONTAMINATION FOLLOWING A TERRORIST CHEMICAL, BIOLOGICAL, AND RADIOLOGICAL ATTACK

JIRI MATOUSEK

Masaryk University Brno, Faculty of Science,
Research Centre for Environmental Chemistry and Ecotoxicology
Kamenice 3, CZ-625 00 Brno, Czech Republic

Abstract

This paper describes the two level system of personal decontamination that was originally developed in the Czech Republic (the former Czechoslovakia) for both the armed forces and civil protection. It is suitable for rescue operations after the release of toxic, contagious agents and radionuclides during both wartime and peacetime operations. The system is suitable for rescue teams as well as for contaminated personnel. The first level of decontamination (primary or immediate) is at the individual level and is considered as a medical countermeasure (in some cases life-saving). The secondary level of decontamination is for non-wounded personnel. It is executed by the Chemical Corps (Civil Protection) at special decontamination sites, while the secondary decontamination of wounded personnel is considered a medical (Health Service) measure and is executed, if necessary, at the entrance of a facility of the medical rescue and evacuation system.

1. Introduction

Decontamination of both rescue personnel and the civilian population following exposure to toxic chemicals, bacteriological (biological) and toxin agents, and radionuclides (CBR) after a terrorist attack is important. For this purpose, standard methods and equipment have been developed, produced, and stockpiled for the armed forces and civil protection. Alternative and improvised methods may also be employed as the situation dictates.

The risk of contamination of naked skin by nerve agents possessing high percutaneous toxicity and a high penetration rate stresses the urgency and effectiveness of primary decontamination. It also demonstrates the need of making this measure available at the lowest possible level, i.e. to the individual. Primary decontamination of naked skin and adjacent parts of clothing should be a medical measure of life-saving importance, akin to the administration of the first-aid antidote. This is valid at any operation where potential physical contact with super-toxic, lethal CW agents is possible. Rescue teams should be equipped with high-quality physical protection methods, first aid equipment, and a capability of primary decontamination. This directly impacts their operational capability.

They also should be equipped with a means of primary decontamination for exposed people, providing them with first aid and additional aid as necessary, within the framework of the medical evacuation system. Immediate on-site diagnostics and corresponding classification is necessary to distribute evacuation streams of exposed wounded and ambulatory people for further treatment consisting of secondary decontamination, with and without medical treatment. This paper presents a system that consists of two-levels of personal decontamination, along with the results of R&D, production, and practice of using these methods of decontamination on individuals.

P. Stopa and Z. Orahovec (eds.), Technology for Combating WMD Terrorism, 103-109.

2. First Decontamination Level: Primary Decontamination

The first post-WW-II generation of primary decontaminants for first aid was the two-solution system manufactured in Czechoslovakia. This system was similar to the Soviet-originated model IPP-51 and was introduced into the Czechoslovak Army with a designation of IPB-60. It was subsequently introduced in a simplified format into the Czechoslovak Civil Defence in the early 1960s. Its designation was **OZB**. The composition of the kit corresponded to then existing CWA, mainly HD and GB, and other classic agents.

This system is based on the subsequent use of two solutions:

Solution No 1 - 15 per cent (w/v) of sodium cresolate in 96 per cent ethyl alcohol.
Solution No 2 - 20 per cent (w/v) of sodium benzene sulphochloroamide (chloramine B) in 82 per cent (v/v) ethyl alcohol in water with 11.6 per cent (w/v) of zinc (II) chloride.

Solution No 1 was designated for decontamination of G-agents and solution No 2 for decontamination of HD and all other CWA. However, if the contaminant was unknown, both solutions could be used in tandem; however, this compromised the performance of these solutions since specific decontamination is more effective.

After the appearance of V-agents, it was necessary to test the effectiveness of this system for the decontamination of these agents. The exact formula of VX is well known and was published in 1974, but suspicion on the probable chemical structure of the V-agents led me to initiate research on impact of V-agents on the whole system of chemical defences (detection, protection, decontamination, first aid etc.). This work was performed at the Czechoslovak NBC-Defence R&D Establishment (nowadays Military Technical Institute of Protection in Brno) and was initiated in the 1963-64 timeframe. The initial research work was performed, using as a standard model substance, the N,N-dimethyl analogue of VX, i.e. O-ethyl S-(NN-dimethyl)aminoethyl methylphosphonothiolate. It was designated as SN-25 or as *Medemo* later in some foreign works. The study of synthesis and other relevant experimental work was performed with a wider group of homologues. The work on the problems of medical protection was started at the then Purkyne Medical Research Institute (nowadays known as the Purkyne Military Medical Academy) in Hradec Králové in the late 1960s.

I commenced research on the decontamination of personnel at the latter institution in 1971. This research was started with the testing of the efficiency of the above-mentioned two-solution system as a point of reference for any further work using biological systems. Testing of percutaneous toxicity and of decontamination efficiency was performed mainly in rats, although some was also performed in guinea pigs and rabbits. For these reasons, standard methods of both contamination and decontamination for liquid and (later developed) solid decontaminants were developed.

I found that the best way to express the decontamination efficiency of a system is the **Decontamination Index** **(DI$_{50}$)**. This expression corresponds to the ratio of LD$_{50}$ p.c. (mg/kg) *with decontamination* (under standard conditions) to the LD$_{50}$ p.c. (mg/kg) *without decontamination*. It is obvious that the higher the value of DI$_{50}$, the higher the decontamination efficiency, while DI$_{50}$ = 1 means no decontamination at all.

To effectively use this index, it was necessary to establish baseline values. Work was performed with several agents on several animal systems. These results are shown in Table 1. They are representative of the percutaneous toxicity of the most important CWA to be tested in decontamination experiments.

To gauge the effectiveness of any new decontamination methods, we established the two-solution system, i.e. IPB-60 and OZB as the baseline. The results of this testing are presented in Table 2. Please note that the decontamination efficiency against the V-agents (VX was tested in the mid-1970s) seems low. If these results (expressed in LD$_{50}$ terms) are translated into the terms of combat contamination density on the skin surface in mg/sq.cm, it is clear, that the decontamination efficiency is insufficient[1]. Moreover, the insufficient rate of the decontaminating reaction even diminishes the protective properties of clothing[2] in the case of decontamination on man[3]. Skin irritancy of human skin also justified the initiation of R&D on new decontamination materials.

Table 1. **Percutaneous toxicity of tested agents LD$_{50}$ p.c. mg/kg[1]**

Agent	Albino rat	Guinea pig	Rabbit
HD	15.82 ± 3.28	---	---
GB	128.40 ± 84.90	$98.49 \, (63.54 - 152.7)$	---
GD	15.77 ± 4.23	22.73 ± 4.98	5.87
SN-25 (*Medemo*)	0.0630 ± 0.0222	$0.809 \, (0.521 - 1.824)$	$0.0572 \, (0.0408 - 0.804)$
VX	0.0132 ± 0.0073	$0.246 \, (0.196 - 0.321)$	$0.0514 \, (0.0210 - 0.115)$

Table 2. **Decontamination efficiency of IPB-60 (OZB) expressed at DI$_{50}$ in albino rats[1]**
(Decontamination 2 min after contamination)

Agent	HD	GB	GD	SN-25 (medemo)	VX
DI$_{50}$	11.4	8.1	23.5	71.7	147

The systematic research proceeded in three main directions: decontaminating solutions (emulsions); gels (pastes); and solid materials with sorption (specifically, chemisorption) properties. The most promising results were obtained with the group of specially treated montmorillonites of domestic origin. The final solution is an acid-treated bentonite (i.e. with active H-centre) enabling the manufacture a simple and very cheap means for primary decontamination. We have designated this mode of employment as "sorption-mechanical"[4,5]. It is based on chemisorption. For optimal results of skin decontamination, some rubbing of this fine powder on the skin surface is recommended. This material, designated with the acronym "DESPRACH" during development (reflecting the Czech for *des*activation and fine powder *prach*), was introduced as a new means for individual decontamination. It is designated as IPB-80 in the Czechoslovak Army[6]. It is a substantial part of the new first aid kit ZPJ-80 in the Czechoslovak Civil Defence[6]. The most important property, the decontamination efficiency, is shown in Table 3.

Table 3. Decontamination efficiency of DESPRACH (IPB-80, ZPJ-80) as DI_{50} in albino rats[4]
(Decontamination 2 min after contamination)

Agent	HD	GB	GD	SN-25 (medemo)	VX
DI_{50}	12.56	3.73	19.03	968	1200

Besides the much higher decontamination effectiveness against the most toxic agents with extremely high percutaneous toxicity (V-agents), the new material meets all of the principal requirements of a good decontaminant. These are:

- Effectiveness against all main types of CWA, i.e. universality
- Speed of decontamination effect
- Non-irritancy on skin
- Non-aggressivity on clothing and equipment
- Simplicity of manipulation and use
- Immediate readiness for use
- Low weight
- Use within a wide temperature range
- Mechanical resistance
- Nearly unlimited shelf life
- Ease of manufacturing
- Accessibility of raw materials
- Extremely low costs.

This system is actually very simple to produce and use. The powder is placed in a hand-operated polyethylene (PE) bottle. The bottle is in the form of a quadrangular prism and it has a small orifice that allows one to spread the powder on the contaminated, naked skin and the adjacent parts of clothing. The bottleneck is then fitted with a screw cap.

This device is an example of an atypically cheap innovation. The price of the new IPB-80 was less than one third as the old IPB-60. The first aid kit for civil defence (ZPJ-80) was a new design.

Table 4. **Personal decontamination kit IPB-80 (Czech and Slovak armies)**

Designation:	Decontamination of naked skin and adjacent parts of clothing
Composition:	Bottle with sorption mechanical decon mean DESPRACH Special soap with complexion agent Gauze tampons Box
Main data:	Box size: 130 x 80 x 40 mm Weight: 180 g (DESPRACH 40 g) Operation temperature: from - 40 to + 50° C

Table 5. **First aid kit ZPJ-80 (Czech and Slovak civil protection)**

Designation:	Personal decontamination Administration of the first-aid antidote against nerve agents Treatment of wounds Treatment of burns
Composition:	Bottle – sorption-mechanical decon mean DESPRACH Bottle – special solution for treatment of burns (and decontamination of eyes) Syringe with the first aid antidote Special soap with complexion agent Bottle – tablets DIKACID for preparing potable water and disinfection Gauze tampons Box

Note: First-aid bandages are outside of the box.

3. Secondary and Higher Decontamination Levels

3.1 SECONDARY DECONTAMINATION OF WOUNDED PERSONNEL

Secondary decontamination of wounded personnel can be performed at any facility of the medical evacuation and rescue system. Every such facility- no matter whether it is stationary or mobile; well-engineered or improvised- typically contains an entrance module (room, tent etc.) and is divided into a dirty (hot or contaminated) part and a clean part. The flow of wounded people proceeds from the dirty part, where the use of decontaminants, complemented with showering, to the clean part. From there patients are discharged to respective treatment modules. This principle has been maintained during several versions of the equipment of the medical rescue system through the last several decades.

For the Medical Services, the original two-solution system (IPB-60) is used in two kits: the **PCHB-60-P** and the **PCHP-60-P**. The facilities at all echelons of the medical rescue and evacuation system are equipped with these kits, starting with the **PCHB-60-P** at the lowest, i.e. company, level. All higher levels are equipped with the latter, taking into account that the **PCHP-60-P** contains ten times more material than the **PCHB-60-P**. This is greatly simplifies the logistical burden.

In case of secondary decontamination, we do not consider the properties of the original IPB-60 as an important disadvantage for two main reasons:

1) The primary decontamination has been already carried out.

2) At the echelons of the medical rescue system, the toxic agent is more likely known. In the selective use of either Solution No 1 or 2, the efficiency is generally considerably higher than necessary, which precludes the subsequent use of both solutions on an individual.

Besides, the sorption-mechanical mean is always available as a mean of choice.

3.2 SECONDARY DECONTAMINATION OF NON-WOUNDED PERSONNEL

Because primary decontamination is generally considered as a medical measure, the Chemical Corps executes the secondary decontamination of non-wounded personnel. This proceeds regularly at sites where decontamination of equipment and other items takes place. Showering is the technique that is primarily used.

The concept for showering of both military and civilian casualties for decontamination purposes is not new. There have been many generations of various mobile methods available, either as a constitutive part or as an accessory to chemical response vehicles. For civilian protection, mostly fixed facilities are utilized that are part of hospitals or hygiene stations. There are also mobile capabilities that utilize equipment from fire brigades.

The personnel decontamination set[7] designated as **SDO** is the latest Czech model of such a facility. Table 6 describes this system. For supplying the SDO, the new Czech decontamination spraying vehicle, **ACHR-90,** is used. It was introduced in the late 1990s[7] and contains a small amount of showering equipment as an accessory. It is designated in the first line for decontamination of the crew with a surplus capability.

Table 6. **Personnel decontamination set SDO**

Designation:	Personnel decontamination, showering	
Composition:	- Three inflated tents	
	- Water system (2 SANNET aggregates)	
	- Heating (cooling) system	
	- Electrical system	
	- Transportation: truck TATRA-815 VN (6WD)	
	- Supply: Decontamination vehicle ACHR-90	
Main data:	Decon capacity	150 persons/hr
	Showering capacity	120 persons/hr
	Continuous operation	10 hrs
	Water temperature	38° C
	Installation time	45 min
	Break-off time	60 min

4. Conclusions

Thanks to systematic R&D, the Czech armed forces as well as civilian population is equipped with an efficient decontamination at all necessary levels. The main emphasis is placed on primary decontamination. It is considered as a medical measure of great importance, in cases of intoxication with super-toxic lethal nerve agents. If these agents are used the administration of antidotes and decontamination are the highest urgencies for the saving of life.

5. References

1. Matousek J.: Sorption-mechanical principle in skin decontamination. In: Sohns T., Voicu V.A. (Eds.): *NBC Risks: Current Capabilities and Future Perspectives of Protection.* Kluwer Academic Publishers, Dordrecht – Boston – London 1999, pp 265-269.

2. Matousek J.: Protective properties of standard clothing against skin penetration of super-toxic lethal chemical warfare agents. In: Sohns T., Voicu V.A. (Eds): *NBC Risks: Current Capabilities and Future Perspectives for Protection.* Kluwer Academic Publishers, Dordrecht – Boston – London 1999, pp 303-310.

3. Matousek J.: Problem of decontamination on man. *6th International Symposium on Protection against CBW Agents.* Stockholm 1998. *Proceedings, vol. 1*, NDRC, Dept. of NBC Defence, Umea 1998, pp 255-256.

4. Matousek J.: Means for decontamination of supertoxic lethal chemicals on human skin. *Symposium on Nuclear, Biological and Chemical Threats in the 21st Century - NBC 2000.* Helsinki- Espoo 2000. *Symposium Proceedings, Research Report No 75,* University of Jyväskylä 2000, pp 216-221.

5. Matousek J.: Universal mean for primary decontamination on the body surface (in Czech), CS Pat. 1403/T.

6. Matousek J., Hodbod J., Sebestik M.: Sorption-mechanical mean for decontamination of CW agents for multiple use (in Czech), CS Pat. 1931/T.

7. Army of the Czech Republic: *Catalogue, NBC Defence & Chemical Support Equipment,* Prague 2001.

FOUR-WEEK-LONG TABUN LOW-LEVEL EXPOSURE IN RATS

MILOS P. STOJILJKOVIC, ZORAN A. MILOVANOVIC,
VESNA KILIBARDA,DUBRAVKO BOKONJIC
National Poison Control Centre & Institute for Scientific Information;
Military Medical Academy, Crnotravska 17

DANKA STEFANOVIC, BILJANA ANTONIJEVIC
Institute of Toxicological Chemistry, Faculty of Pharmacy, Vojvode Stepe 450;
YU-11000Belgrade, Federal Republic of Yugoslavia

Abstract

Prolonged administration of sublethal doses of organophosphorus cholinesterase inhibitors results in adaptation to their toxicity. In order to investigate this phenomenon, we exposed rats to 0.2, 0.3 or 0.4 LD_{50} of tabun *sc* daily during four weeks. AChE activities in erythrocytes, diaphragm and brain were inhibited dose-dependently after days 7 and 14 of the study and started to recover thereafter, except after 0.4 LD_{50}. Tabun 0.3 LD_{50} decreased body weight gain, food and water consumption during the first two weeks of the study. Spontaneous locomotor activity was significantly increased in the same interval and decreased thereafter. These findings support the assumption that both the biochemical and receptor mechanisms are responsible for the occurrence of tolerance to tabun in rats.

1. Introduction

There is a plethora of the experimental and clinical studies dealing with the consequences of the poisonings with organophosphorus compounds (OPCs), including OP insecticides (OPIs) and nerve agents tabun, sarin, soman and VX. At the same time, the corresponding data on the effects of subacute low-level exposure to OPCs are rather scarce (1). The majority of the presently known publications on the subjects are four or more decades old and they are dealing with subacute poisoning with OPIs, such as parathion (2), octamethyl pyrophosphoramide (3), chlorthion, tetrapropyl dithionopyrophosphate, malathion (4), dipterex (5), systox (6) and di-syston (7). After proving that repeated administration of a meant-to-be-nerve agent diisopropylfluorophosphate (DFP) leads to development of tolerance (8), the first experimental study addressing the same issue in a similar manner, but employing real nerve agents instead of OPIs, was the one on tabun, sarin and soman subacute toxicity in rats (9). The reason for this disparity between the research in the OPI and nerve agent field was the fact that interest in the effects of OPCs after prolonged low-level exposure originated from the industrial hygiene. In fact, the stimulus for the investigations came from the notion that those individuals that are engaged in the manufacture or use of the OPIs could receive and survive repeated exposure to small doses. It was much later, when the military toxicologists started to share this initially purely civilian interest in the subacute toxicity of the specific OPCs, i.e. nerve agents (9-15).

Tabun was the first nerve agent to be synthesised by Dr. Gerhard Schrader of I.G. Farbenindustrie, on 23 December 1936 (16). The aim of this study was to try to shed some more light on the effects of the subacute effects of this potent irreversible inhibitor of acetylcholinesterase (AChE), when administered on a daily basis and to reveal the underlying mechanisms of tolerance to tabun.

2. Methods

Male Wistar rats were used throughout the experiment. The median lethal dose (LD_{50}) of tabun was 150 μg/kg subcutaneously (*sc*). Animals were exposed to 0.2, 0.3 or 0.4 LD_{50} of tabun *sc* daily over a four-week period during

P. Stopa and Z. Orahovec (eds.), Technology for Combating WMD Terrorism, 111-117.

which a number and time of fatalities was registered. Groups of control and tabun-poisoned animals (n=4) were sacrificed on days 7, 14, 21 and 28 in order to obtain tissue samples for biochemical analyses. For these purposes AChE activities in brain, diaphragm and erythrocytes were analysed spectrophotometrically (17). In the group of rats treated with 0.3 LD_{50} of tabun, body weight gain, food and water consumption (18) and spontaneous locomotor activity - or SLA (19) were monitored on a weekly basis.

3. Results

All the rats treated on a daily basis with 0.2 LD_{50} of tabun survived four weeks. Cumulative mortality in the 0.3 LD_{50} group was 17% - the first animal died on the day 12 and the last one on the day 20 of the study. In the 0.4 LD_{50} group, a total of 88% died, between the treatment days 6 and 20 (Figure 1).

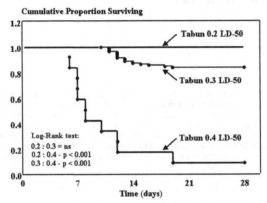

Figure 1. Cumulative mortality rates in rats treated with tabun 0.2, 0.3 or 0.4 LD_{50} daily.

AChE activities in all the tissues investigated and especially in the brain was dose-dependently inhibited with tabun. The sharpest decrease in the enzyme activities was found after days 7 and 14 of the study. In two lower dose groups gradual recovery of the enzyme activities was noticed during the last two weeks (Figure 2).

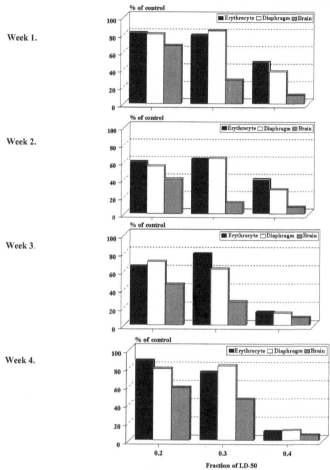

Figure 2. Activities of AChE in erythrocytes, diaphragm and brain in rats treated with tabun 0.2, 0.3 or 0.4 LD_{50} daily.

The three doses of tabun dose-dependently decreased body weight gain during the first two weeks. During the remaining period, body weight gain approached the level in the control group. Tabun in a daily dose of 0.3 LD_{50} decreased body weight gain, food and water consumption especially during the first two weeks of the study. SLA was increased in the same interval and decreased thereafter (Figure 3).

114

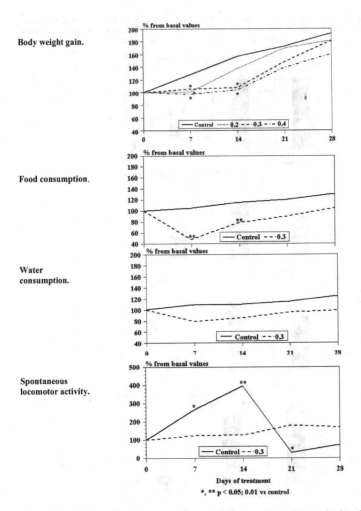

Body weight gain.

Food consumption.

Water consumption.

Spontaneous locomotor activity.

Days of treatment

*, ** p < 0.05; 0.01 vs control

Figure 3. Nutritional and behavioural parameters in rats poisoned sublethally with tabun during 4 weeks.

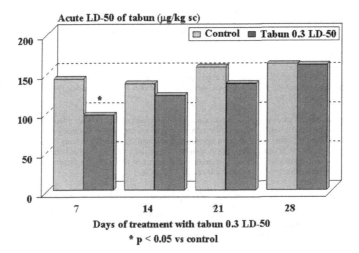

Figure 4. Acute LD_{50} values of tabun in naïve rats and animals treated with tabun 0.3 LD_{50} daily.

Daily treatment of rats with 0.3 LD_{50} of tabun significantly increased acute tabun toxicity only after the first week and gradually approached the control value by the week 4 of the study (Figure 4).

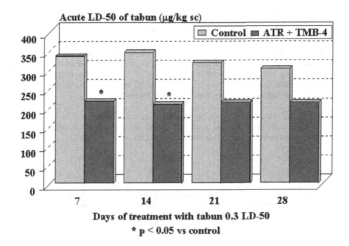

Figure 5. Acute LD_{50} values of tabun in naïve rats and animals treated with tabun 0.3 LD_{50} daily. All the rats received atropine 10 mg/kg and TMB-4 10 mg/kg *im*, immediately after acute tabun administration.

When both the control and the tabun-treated rats were poisoned acutely with tabun and immediately thereafter treated with atropine and TMB-4, the acute tabun LD_{50} values remained constantly around 250 µg/kg, making the difference to the naïve rats significant only after the first two weeks (Figure 5).

4. Discussion

Tabun was chosen for this study, because it is a very potent irreversible AChE inhibitor and because its equitoxic doses exert the strongest and the fastest tolerance, compared to the ones of sarin or soman (9). This was the reason why the study was set to last for four weeks.

Although some of the rats received in four weeks tabun in a dose that exceeds 11 acute LD_{50}s, which equals more than 8 absolute acute lethal doses of this nerve agent, they survived the experiment. Comparable levels of tolerance - to more than 14 times the acute LD_{50} of nerve agents were obtained by the other authors (9). The results of the present study that lasted for 28 days, along with the similar ones published earlier (9) that lasted for 85 days, clearly indicate that time is a crucial factor for the occurrence of tolerance to OPCs. At the same time, they explain the results obtained in some shorter experiments that lasted for 11 days only (20-22).

Among the general toxic parameters monitored, body weight gain seems to be the most sensitive indicator of the occurrence of tolerance to nerve agents (9). Its decrease always accompanies the sharp decrease in the tissue AChE activities that could be observed during the first two weeks. On the other hand, during the remaining two weeks of the experiment, when tolerance has already occurred, relative plateaus in tissue AChE activities parallels normalisation of the body weight gain and food consumption.

Changes in SLA observed deserve some specific comments. Although it is generally well known that the anticholinesterase agents - both organophosphates and carbamates - decrease the whole-body locomotion (23, 24), our results for the first 14 days of daily treatment with 0.3 LD_{50} of tabun are quite opposite. These at first sight maybe paradoxical results are however in accordance with the ones obtained by D'Mello (19). In his experiment on marmosets a small dose of sarin of 7.5 µg/kg intramuscularly (*im*) increased their whole-body locomotion three times. Moreover, 12.5 µg/kg *im* of sarin returned the number of movement to the control level, while the dose of 17.5 µg/kg *im* decreased it by some 50%. Therefore, nerve agents, administered in smaller doses - up to 0.4 LD_{50} - stimulate SLA, while higher doses have the opposite effect (1). This phenomenon is clearly shown after sarin (25, 26) and soman (27, 28) and repeated dosing with fenitrothion (29) at doses in a range 0.3-0.8 LD_{50}.

Generally speaking, spontaneous regeneration of tabun-inhibited AChE in various tissues is slow. Its regeneration half-times in whole blood, diaphragm and intercostal muscles amount 6.6, 3.6 and 5.9 days, while the same parameter ranges among the brain regions from 7.4 days in basal ganglia to 11.9 days in cortex (30). Notwithstanding these data, it is obvious that in our experiment 0.2 and 0.3, but not 0.4 LD_{50} of tabun daily assured gradual return towards the initial values after weeks 3 and 4 of the study.

Although at least part of the explanation for the occurrence of tolerance to tabun in rats should be found in its metabolism (31), there is no doubt that the other part should rest on the grounds of desensitisation and down-regulation cholinergic and probably some other brain receptors (32).

5. Conclusion

Data on the changes in AChE activity support the assumption that a specific biochemical adaptation to the presence of an organophosphorus cholinesterase inhibitor occurs in the organism of experimental rodents, especially after the week 2 of the experiment. The receptor mechanisms of this adaptation cannot be ruled out, too.

6. References

1. Du Bois, K.P. (1963) Toxicological evaluation of the anticholinesterase agents. In: Koelle, G.B. (editor) Cholinesterases and anticholinesterase agents, Springer-Verlag, Berlin, pp. 833-859.

2. Du Bois, K.B. et al. (1949) J. Pharmacol. Exp. Ther. 95, 79-91.

3. Du Bois, K.B. and Coon, J.M. (1950) J Pharmacol. Exp. Ther. 99, 376-393.

4. Du Bois, K.B. et al. (1953) A.M.A. Arch. Industr. Hyg. 8, 350-358.

5. Du Bois, K.B. and Cotter, G.J. (1955) A.M.A. Arch. Industr. Hlth 11, 53-60.

6. Barnes, J.M. and Denz, F.A. (1954) Br. J. Industr. Med. 11, 11-19.

7. Bombinski, T.J. and Du Bois, K.P. (1958) A.M.A. Arch. Industr. Hlth 17, 192-199.

8. Russell, R.W. et al. (1975) J. Pharmacol. Exp. Ther. 192, 73-85.

9. Dulaney, M.D. Jr. et al. (1985) Acta Pharmacol. Toxicol. 57, 234-241.

10. Bajgar, J. (1992) Br. J. Industr. Med. 49, 648-653.

11. Husain, K. et al. (1993) J. Appl. Toxicol. 13, 43-145.

12. Haley, W. and Kurt, T.L. (1997) J. Am. Med. Assoc. 277, 231-237.

13. Brown, M.A. and Kelley, A.B. (1998) 18, 393-408.

14. Ray, D.E. (1998) Toxicol. Lett. 102-103, 527-533.

15. Kassa, J. et al. (2001) Pharmacol. Toxicol. 88, 209-212.

16. Stojiljkovic, M.P. (1997) Prophylaxis of poisoning with soman, Andrejevic Foundation, Belgrade.

17. Ellman, G.L. et al. (1961) Biochem. Pharmacol. 7, 88-94.

18. Glow, P.H. et al. (1966) J. Comp. Physiol. Psychol. 61, 295-299.

19. D'Mello, G.D. (1992) Neurobehavioural toxicology of anticholinesterases. In: Ballantyne, B. et al. (editors). Clinical and experimental toxicology of organophosphates and carbamates, Butterworth-Heinemann, Oxford, pp. 61-74.

20. Sterri, S.H. et al. (1980) Acta Pharmacol. Toxicol. 46, 1-7.

21. Sterri, S.H. et al. (1981) Acta Pharmacol. Toxicol. 49, 8-13.

22. Sterri, S.H. et al. (1982) Acta Pharmacol. Toxicol. 50, 326-331.

23. Koehn, G.L. and Karczmar, A.G. (1978) Prog. Neuro-Psychopharmacol. 2, 169-177.

24. Romano, J.A. Jr. and Landauer, M.R. (1986) Fundam. Appl. Toxicol. 6, 62-68.

25. D'Mello, G.D. and Duffy, E.A.M. (1985) Fundam. Appl. Toxicol. 5, S169-S174.

26. Landauer, M.R. and Romano, J.A. (1984) Neurobehav. Toxicol. Teratol. 6, 239-243.

27. Raffaele K. et al. (1987) Pharmacol. Biochem. Behav. 27, 407-412.

28. Romano, J.A. et al. (1985) Proc. 5th Annu. Chem. Def. Biosci. Rev., Columbia, USA, 29-31 May 1985.

29. Rondeau, D.B. et al. (1981) Neurobehav. Toxicol. Teratol. 3, 313-319.

30. Stojiljkovic, M.P. et al. (2001) Proc. 7th Int. Symp. Protect. Chem. Biol. Warfare Agents, Stockholm, Sweden, 15-19 June 2001.

31. Stevens, J.T. et al. (1972) J. Pharmacol. Exp. Ther. 181, 576-583.

32. Sivam, S.P. et al. (1983) J. Neurochem. 40, 1414-1422.

PARADIGM SHIFT IN DEVELOPING COUNTERMEASURES TO BIOLOGICAL AND CHEMICAL THREATS

WALTER L. ZIELINSKI AND STEVEN KORNGUTH
Biological and Chemical Defense
Institute for Advanced Technology, The University of Texas at Austin

Abstract

Developments in the continental United States (dissemination of Anthrax-contaminated items through the U.S. mail in October 2001) and on the international scene (emergence of Congo-Crimean hemorrhagic fever in the Afghanistan-Pakistan border region) have alerted the defense community to the real threat of biological and chemical agents to civilian and military populations. Current strategies have had a stove-piped appearance, with efforts directed towards the development of individualized communications systems, separate detectors for each threat agent, tradition-based vaccines, and antibiotics for post-event care delivery. What is needed is an integrated sensor alert system that can selectively and simultaneously detect all threat agents with little to no false positive/negative events, a seamlessly integrated communications network capability that enables the conversion of data to actionable information, and novel pre- and post-event treatments. New technologies required for an effective response to a biological or chemical attack include: multi-array sensors for threat agents; sensors for validating signatures of the host response to infection; a telemedicine system to deliver post-event care for up to 20,000 victims of a biological or chemical strike; an intelligent software network that permits meaningful resource allocations and redundant pathways that survive a catastrophic attack; and novel vaccines and restricted access antivirals/antibacterials to reduce the emergence of drug resistant strains pre- and post-event. Multi-array sensor platforms and alert systems must permit rapid screening of many samples for the presence of signature molecules and be small and lightweight with low power requirements. Genomic sequences of threat agents naturally present in different global regions must be determined to permit attribution of an attack to the source of an agent. A communications system that allows for the integration of information on emerging disease, resource allocation inventory, public safety coordination, perimeter management and extended telemedicine care is yet to be realized. Databases that provide information on the normal distribution of naturally occurring threat agents and the normal incidence of diseases in communities must be established. These databases should be integrated into a sensor and communication system that will provide an early alert of a biological or chemical attack, and optimize consequence management actions to neutralize the impact of the attack. This sensor and communication strategy is critical to sustaining normal health, commerce and international activities.

1. Introduction

The overarching objectives in developing effective countermeasures to biological and chemical threats are protecting the U.S. Defense community and citizenry from such threats, and developing agile responses to unanticipated events – considering that terrorists frequently do the unexpected. The need for protection against and responses to biological and chemical threats has been strikingly demonstrated by the use of anthrax-contaminated letters that were sent through the US mails in October 2001. That attack resulted in human illness, the loss of life and contamination of federal office buildings in Washington, DC. The incidence of Congo-Crimean hemorrhagic fever in Afghanistan and Pakistan, areas where US forces have been heavily deployed increased this need. The potential threat posed by a major release of a contagious biological agent (e.g., smallpox) has been of growing concern at all levels, from administration leaders to the US public. This article outlines and discusses directions that are critically needed if we are to be fully capable of sensing, preventing and responding to biological and chemical threats.

P. Stopa and Z. Orahovec (eds.), Technology for Combating WMD Terrorism, 119-124.
© 2004 *Kluwer Academic Publishers. Printed in the Netherlands.*

2. New Paradigm

The current paradigm addresses biological and chemical terrorist threats in a vertical (stove-piped) response. In the arena of developing sensors for the detection of biological agents, the paradigm has been to develop separate detectors for each agent. Interactive networked communications systems that are critical to an effective response to a major terrorist event are lacking. For major therapeutic responses, reliance has been on the production of vaccines, antibiotics and antivirals using 40-year-old technology protocols. These require extensive development times before they become available for human use, and undesired side effects commonly result from vaccines produced by these protocols. Although improvements in emergency medical response and hospital equipment have been achieved, the ability of any US community to manage an outbreak of infectious disease affecting >10,000 people is lacking.

The new paradigm utilizes a network centric integrated sensor alert system that can detect all threat agents simultaneously, coupled with a seamlessly integrated communications capability that converts sensor data to actionable information. For this to be effective, the sensor system must yield minimal false positive and false negative results. The new paradigm likewise incorporates novel pre- and post-event treatment capabilities.

What are our national needs in this new paradigm? We need detectors for biological agents that infect both people and the human food chain, including livestock and edible crops. The agents of concern include many on the Australia Group biological/toxin warfare agents list. We need sensitive alert systems that have a capability for the early detection of host responses to infection. We need a new generation of effective therapeutics, including vaccines, antibiotics and antivirals. We need an intelligent communications network that can provide accurate information to decision makers. We need a telemedicine infrastructure capable of responding rapidly and effectively to a catastrophic biological incident (>10,000 casualties). In addition, to minimize morbidity and mortality and optimize containment for disease spread, we need a biosurveillance system based on archival health databases, statistical models, and data mining strategies that can provide an early alert to a disease outbreak.

3. Where are we on the scale of reaching this new paradigm?

In the sensors area, the genome of most biological threat agents has been sequenced and the signatures of toxins described. Novel multi-array sensor-platform systems may use this information in selectively detecting these agents. In the therapeutics area, researchers are working towards identifying critical antigenic epitopes of these agents. New therapeutics can emerge that have an antigen binding capacity that can be significantly greater than antigen-cell receptor binding, resulting in the potential for agent neutralization. In the new paradigm, novel methods are foreseen for the development of new drugs and DNA based vaccines. Restricted access antivirals/antibacterials will need to be developed to reduce the emergence of drug resistant strains pre- and post-event. A significant development in our program (ref. 1) has been the novel design and production of an antibody that has a high binding affinity ($K_d < 10^{-10}$) for the anthrax PA antigen that binds greater than 20 times stronger than any antibody to date. In tests with experimental rodents in a controlled facility, administration of the *Bacillus anthracis* PA antigen to the animals resulted in 100% fatalities, whereas the co-administration of the newly developed antibody against the PA antigen resulted in 100% survival (ref. 2). Research is being conducted to determine unique nucleic acid sequences (pathogenicity islands) in the genome of a bacterial threat agent. This information is being used to develop multiplexed assay systems that can detect selected agents simultaneously. By quickly screening for multiple pathogenicity island sequences, end-users will have the capability to detect the first signs of a bioattack without requiring screening for a particular organism.

In the communications area, researchers are developing 'belief maintenance' software to provide decision makers some estimate of the validity of incoming data. An estimate of information credibility is critical to effective decision-making in crisis situations when one must rely on an 80 percent solution (i.e., 80% of needed information is available). Waiting for a 100 percent solution could have a catastrophic impact on response effectiveness.

In the area of telecommunications, researchers are developing the means to provide effective medical triage to victims in a contaminated 'hot zone'. With suitable telecommunications, physicians and other medical experts at remote locations can provide telemedicine information and support care delivery to personnel in the hot zone. Prior training is required for these personnel, who could range from physicians to local citizens. In a hot zone, local medical staff will need non-medical civilians - or soldiers, in a military hot zone – to provide support to 'back-fill' the overburdened local community.

In the area of detecting an outbreak of illness, researchers are developing biosurveillance systems to serve as an early warning to evolving disease. For these systems, a variety of databases are being developed that are health related. Examples of these databases include school absenteeism data, over-the counter drug sales, hospital emergency clinic data and archival data on the incidence of diseases in different geographic regions, CONUS and OCONUS. Each database must be statistically characterized regarding parameters such as variance, confidence intervals, seasonality, etc., and be integrated into validated predictive models. Once the reference databases are in place and suitably modeled, statistically significant departures from baseline values can be detected and transmitted in real time, together with information from a threat agent sensor network, to decision makers through intelligent communications systems.

3.1 TECHNOLOGY GAPS

As indicated above, a number of critical technology gaps exist that must be addressed if we are to recognize, prevent and minimize the effect of biological and chemical agents. These gaps include: deficiencies in the availability of multi-agent sensors and platforms; critical reagents; the capability for large-scale production of high purity vaccines, antibiotics and antivirals; the ability to treat a large number of people for a prolonged period in a biological hot zone; and the existence of archival biosurveillance databases and intelligent and secure communications networks. Some approaches to address these gaps include the use of autonomous (e.g., cell phone-based) microelectronic detectors for the transmission of data on agent exposure, the development of novel antibodies, antibiotics and antivirals to manage disease outbreaks, the development of a national medical infrastructure to triage a large community hit by a biological strike, and the establishment of global surveillance systems for emergent diseases (e.g., West Nile Fever, Congo-Crimean Hemorrhagic Fever).

3.2 RESEARCH AREAS

Because of the broad scope of our nation's technology needs for preventing and minimizing biological and chemical (B/C) threats, a number of research areas have been identified as critical. These include: the scientific validation that a B/C incident has occurred (requisite tools/capabilities include situation awareness systems, sensors and signatures); the availability of medical countermeasures (vaccines, pharmaceuticals and medical transport); and a highly effective communications network for the secure transmission of accurate data to decision makers for appropriate actions.

3.2.1 Sensors research - For effective sensors, a variety of materials are being developed. These include effective high-affinity binders of B/C threat agents such as antibodies, cDNA gene probes, polynucleotide aptamers, and combinatorial chemicals. Using phage display methods, antibody fragments can be selected that have a high affinity for agents such as anthrax toxin and brucella. Another binding system that has been examined uses polynucleotide aptamers about 31 nucleotides long that have good binding affinity to ricin toxin. These sensor materials require opto/electronic transduction platforms. Sensor platform research currently is being focused on micro-electro-mechanical systems (MEMS) devices, microelectronics technology, microfluidics (laboratory-on-a-chip), DNA/protein microarrays and transduction devices. Current efforts also are being directed towards the development of multi-plexed sensor systems that detect a number of biological threat agents of concern. For military application, it is essential that sensor systems can rapidly detect and identify agents present in samples, and that they be small and have low power requirements.

3.2.2 Therapeutics research - With current approaches, the development-to-market of new vaccines, antibiotics and antivirals is in the order of 5-10 years. A paradigm shift to newer culture and DNA-based technology is needed if we are to have an effective response to a major biological or chemical attack. Current estimates for the availability of vaccines and antibiotics/antivirals to specific threat agents using new technology and expedited approval is in the order of 3 years.

3.2.3 Communications research – While current computer/informatics research includes the development of telecommunications assets, a critical need in the communications area is the development of seamless integrated communications networks. These network centric systems enable the conversion of data to actionable information. Research studies are being conducted to provide intelligent agent software designs for such communications systems. This will enable an enhanced accuracy in critical decision-making and resource allocations. The integrated system must have redundant pathways that can survive a catastrophic attack. The communications system will be capable of integrating data on an emerging threat, public safety coordination, and perimeter management for responsible situation action officials.

3.2.4 Telemedicine needs - Telemedicine capabilities can aid in the delivery of post-event care to 10,000-20,000 victims of a biological strike in a densely populated area in the US for a period of 24 hours a day, seven days a week, for several months. In the event of a smallpox attack in which 10,000 people develop clinical symptoms of infection within 7-10 days following exposure, local hospitals and medical response capabilities would be overwhelmed. Telemedicine allows physicians at a remote location from the hot zone to provide medical support via telemedicine capabilities (visual, audio, haptic) to aid local physicians in treating hot zone patients. A treatment level of 50 patients per day per physician would require 200 physicians to provide telemedicine care for 10,000 patients. Each physician would require telemedicine devices; hence 200 telemedicine devices would be required at the remote location, and a similar number in the hot zone. A national telemedicine system could include the establishment of approximately eight telemedicine response centers nationally, interconnected via satellite to telepresence and telemedicine/robotic systems. The remote care capability reduces the likelihood of the dissemination of disease to physicians and communities in which the physicians reside. It would also retain health care delivery in communities providing health care back-fill.

4. Command Center

We envision a Command Center for coordinating the delivery of actionable information to decision makers. Figure 1 illustrates such a flow chart. In threat situations, data developed from sensor arrays, surveillance systems, and therapeutics inventories can be electronically encrypted and transmitted via intelligent communications networks to decision makers for appropriate actions (Figure 2). The Command Center is comprised of a Technology component, a Warrior component and an executive authority component. The Technology component includes sensor and communications experts. The Warrior component includes medical care experts (doctors, hospitals, State Boards of Health, the Centers for Disease Control and Prevention (CDC), etc.), the National Guard and the military. The executive authority includes the President and his designates and the Department of Defense. Based on the nature of the incoming information, the Command Center decision makers assess the information that has been compiled in real time and determine the course of action(s). Such actions may require the utilization of personnel assets or the relay of information to appropriate officials and agencies.

5. Acknowledgements

The authors express their appreciation to support received from our sponsors at DTRA (John Oprandy), SBCCOM (James Valdes and William Lagna), and ARL (Edward Schmidt). The authors also acknowledge their appreciation to Robert Chin for development of the concepts and constructs in Figures 1 and 2.

The research reported in this document was conducted under DAAD13-02-C-0079 with SBCCOM and DDAD17-01-D-0001 with the U.S. Army Research Laboratory. The views and conclusions contained in this document are those of the authors and should not be interpreted as representing the official policies or position, either expressed or implied, of SBCCOM, the U.S. Army Research Laboratory or the U.S. Government.

Command Center

Figure 1. Command Center

Command Center Elements

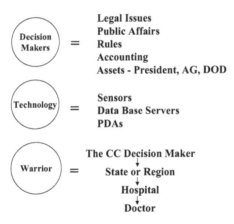

Figure 2. Command Center Elements

6. References

1. Gaunt, D.L., Kornguth, S.E., *The University of Texas Biological and Chemical Countermeasures Program.* Journal of the ANSER Institute for Homeland Security, September 28, 2001.

2. Maynard, J.A., Maassen, C.B.M., Leppla, S.H., Brasky, K., Patterson, J.L., Iverson, B.L., Georgiou, G., *Protection against anthrax toxin by recombinant antibody fragments correlates with antigen activity.* Nature Biotechnology, 20, 597-601, 2002.

SYSTEMS INTEGRATION

PREFACE

Once technologies have been identified to detect, protect, and mitigate the effects of a WMD attack, these items and their supporting personnel and equipment must be assembled in to a response system.

Countries may need to strengthen laws, develop new agencies, and increase cooperation with neighboring states. An example of how one small country responded to the increase threat of terrorism is discussed.

Risk analysis may provide a paradigm for planners to follow in assessing the types of equipment that one might need to mitigate the effects of a WMD attack. A practical example of this approach is presented whereby this risk-based model can be used by planners to determine the suitability of gas masks in a response plan.

Epidemiological monitoring is also a key component of a response to WMD, particularly the biological component. This can play a key role in recognizing an attack, particularly if it is unannounced.

The components of the response eventually need to integrated into a system. There needs to be a command and control structure, situational awareness, and training for an effective system. These components need to be managed. Currently, there are several computer-based tools that can be employed for this purpose. An example of one such tool, and how it can be implemented, are presented.

DEFENSE AGAINST TERRORISM

ALFRED MADHI
Ministry of Defense
Tirana, Albania

1. Legislative Approaches

Albania, as a factor of peace and stability in the region, does not support terrorist acts in all of its forms, and will continue to devote its efforts for the strengthening of regional and European security. In this context, the Albanian Parliament has approved Law No 7, dated 10 October 2001, entitled "Albania is a member of the coalition against worldwide terrorism", which states *"The Ministry of Justice, the Ministry of Public Order, the Ministry of Defense, and the National Intelligence Service need to review the related legislation and make further improvements for fighting against the terrorism"*.

The Republic of Albania has aligned itself with the International Coalition against Terrorism. It has taken appropriate measures in fulfilling the requirements of the international agreements to prosecute or extradite terrorists, as well as prevent and combat terrorism. These measures include:

- the establishment of a Regional Center for fighting against illicit trafficking- drugs, smuggling, and illegal emigration.

- increasing border control, in order to prevent whatever international terrorist elements from entering the country.

- restructuring of both the Interior Police Force and the Military Police Force.

- establishment of anti-terrorist and drug units within the Ministry of Interior.

- fulfilling the legal framework on antiterrorism and drug traffickers, as described in Article No. 230 of the Penal Code of the Republic of Albania, which states: "Performance of violent acts against life and health of people, kidnapping and their freedom, mass frightening and insecurity in the public order, are condemned to jail, for not less than 15 years, life imprisonment, or to death";

2. Regional Cooperation

The Ministry of Public Order dedicates special importance to the regional cooperation, in terms of coordination of activities on the fight against crime and to exchange experiences of the agencies involved. In cooperation with the Ministry of the Interior of Italy, to aid in the fight against criminality, joint investigations on criminal matters have been conducted, such as drug trafficking, weapon smuggling, human trafficking, stolen cars trafficking, followed by arrests and extradition in both countries.

Albania and Greece have established an agreement, that facilitates cooperation focused on the trafficking of human beings, drugs, illegal migration, border management, etc. The cooperation with FYROM consists on signing legal instruments to discuss the trans-border issues, on preventing illegal activities.

The cooperation with Montenegro has been effective for fighting against human trafficking, smuggling of goods between the borders, and stolen vehicle transport.

Albania and UNMIK have established cooperation's during the last two years. It has been focused in the exchange of information, resulting in the arrest of the perpetrators, identifying the stolen vehicles, and verification of documents.

P. Stopa and Z. Orahovec (eds.), Technology for Combating WMD Terrorism, 129-130.
© 2004 *Kluwer Academic Publishers. Printed in the Netherlands.*

Close cooperation with Interpol and police force of other countries is established to fight against terrorism. The establishment of the Leadership and Civic Education Branch in both Ministry of Defense and Ministry of Public Order has been conducive to the strengthening of the links between police force and the people. Albania full supports the campaign against international terrorism and made available all its airspace, port and airport facilities to the international joint struggle.

Albania is a signatory of a number of international conventions against terrorism, such as:

- The European Convention on the suppression of terrorism ratified by the Albanian parliament on 21.09.2000

- The European Convention on Extradition with the two additional protocols, ratified on 19 May 1998.

- The European Convention on Mutual Assistance in Criminal Matters, ratified on 4 April 2000.

Recently, the Albanian Government has taken appropriate measures and is contributing directly in the fight against terrorism, such as:

- The participation with one platoon (30 soldiers) in Afghanistan with the International Security Forces (ISAF)

- The restructure of the Albanian Armed Forces foresees a strong support in the fight against terrorism, this in cooperation with the Ministry of Interior and other institutions.

- The fight against illicit trafficking is one of the main priorities of the Government. It considers it as the main support in combating terrorism. The Government is exercising control against the human trafficking in the country and has destroyed all the dinghies in the coast so far.

- The Naval Forces are being developed as a coastal guard service.

- The Air and Naval Observation Systems will be developed (adopted) in a way of meeting the needs to fight against terrorism.

Editor's Note: This paper was included as an example of how countries strengthened their internal laws and also formed regional cooperative agreements with their neighbors, largely in response to the events of September 11, 2001. These efforts, and those of other countries, show that the world is committed to stop terrorist acts of any kind.

BASIC REQUIREMENTS FOR THE DEVELOPMENT OF AN INTEGRATED MANAGEMENT AND TECHNOLOGY SYSTEM FOR COMBATING WMD TERRORISM

BOZIDAR STOJANOVIC
IAUS
Bulevar kralja Aleksandra 73/II
11000 Belgrade, Serbia

1. Introduction

Chemical, biological agents and weapons, as well as nuclear and radiological weapons (CBR), were regarded as the most dangerous weapon systems during 20'th century. Today the primary Weapons of Mass Destruction (WMD) threat is no longer directed exclusively against military troops, but particularly against civilian population. Industrial chemicals provide additional dimension to the threat of major accidents as well as large-scale fires. On a small scale, both production and handling of CBR agents implies that many countries or terrorist organizations can obtain such weapons. Further, terrorist organizations often inflict their damages to specific targets as opposed to full-scale conflicts.

In the military WMD arena the threat and defense are well defined. Standardized military biological and chemical weapons (BCW) lost their strategic importance. Their use in the future should be considered as low probable events at tactical level. Large industrial chemical accidents provoked by intention or by terrorist attack are events of medium probability, but, nevertheless, could cover large areas.

The most vulnerable group in all scenarios of CBR attacks is unprotected civilian population. The area of chemical terrorism is strongly influenced by two most well known attacks using sarin were in Matsumoto City (1994), and the Tokyo subway attack (1995), by the religious sect Aum Shinrikyo [1]. These events had a tremendous impact on civil defense activities, especially on the counterterrorism programs in many countries.

Despite the fact that terrorists have so far largely limited themselves to low-tech BCW attacks, the possibility that they could graduate to the use of weapons of mass destruction is too frightening to be ignored. Some of US experts and institutions agree that the likelihood of an event can no longer be ignored , e.g. "It is not a matter of IF, but rather WHEN such an event will occur"[2].

Since 1995, the United States has allocated enormous resources to combating WMD terrorism. Numerous government programs have been created in an effort to prevent and deter terrorism or to mitigate the effects of a major attack, should it occur. The reason is that much effort to improve the capabilities to combat the threat of CBW terrorism, and to manage the consequences should an attack ever occur, has been based on worst-case scenarios. These worst-case scenarios tend to be shaped by perceptions of vulnerabilities and the availability of technology for response to a CBW attack [3].

Critics of the US large counter chemical and bioterrorism programs say these are ill-focused, throwing money in all directions without proper assessment of the threat. In fact, this could even increase the risk of a BCW attack. They consider that considerably less effort has been made to fully understand the nature of the WMD threat, and as result of such an approach, the significance of the source and concentration of the contaminant, as well as the potential for a contaminant migration/exposure pathway and vulnerability of potentially affected groups, usually have been overestimated [4]. The media attention on US vulnerabilities is likely to stimulate the interest of other states and terrorists in such weapons of mass destruction [5].

On the other hand, according to the historical record, attacks with chemical and biological weapons are strikingly infrequent and the number of fatalities and casualties are far lower than those caused by conventional explosives.

Europe is perhaps not free from groups with NBC-terrorism as a weapon to gain publicity, but probably such groups are rare. Comparing Europe with other parts of the world it can be argued that other parts, the Middle East and USA, are more likely to be the target for terrorism and thus need to put more effort on actions to minimize such risks [6].

P. Stopa and Z. Orahovec (eds.), Technology for Combating WMD Terrorism, 131-143.
© 2004 *Kluwer Academic Publishers. Printed in the Netherlands.*

In spite of such estimate certain European countries consider reasonable countermeasures and preparedness measures based on a careful assessment of the nature of the real threat.

In contrast to the military, a process to develop clearly defined requirements does not yet exist for civilian populations. Civilian organizations have different needs and resources than the military. The potential for a terrorist attack against civilians poses many challenges that require different responses than the military.

This paper will analyze key characteristics of hazard assessment and risk assessment and management methodologies. There is proposed a new approach to CBR terrorist incident risk assessment and management as well as a new approach for defining basic requirements for protective measures based on risk assessment and the dose-effect relationship.

2. WMD Threat Response Framework Based on Hazard Assessment

Hazard assessment is the most commonly used methodology for analyzing the effects of some agent or substance on the human health and on the natural environment. Hazard is commonly defined as "a threat which could cause harm", and could be simply expressed as:

Hazard = Exposure x Effects.

Much of the analysis of the WMD terrorism threat has so far been directed towards circumscribing the hazard imposed by chemical/biological/radiological weapons. Estimation of a threat involves identification of the adverse effects which a substance or agent has an inherent capacity to cause. The hazard identification process is based on a model of CBR terrorist incident consisted from three categories: potential actors in a terrorist incident, MD agents and objects in target areas (Figure 1).

For example, agents could include various types of chemical, biological, radiological weapons or materials. Actors in a potential terrorist attack could be from traditional political groups, religious radical groups, individuals acting alone, etc. Among the target category the most vulnerable probably are various subcategories of civilian populations, because these groups have low potential to modulate its response to stressors over time and space.

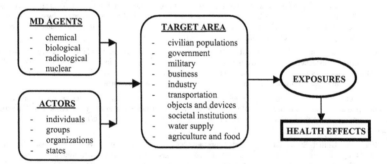

Figure 1 - Elements of a CBR terrorist incident model

Some combinations and the interaction of acting factors on a target create effects or consequences. The most current consequence projections are based on a simple equation: increased toxicity or pathogenicity equals high casualties. For example, calculations based on the theoretically most hazardous agents led to following estimations: (1) the effective dosage of an CWR agent is say 1 milligram, if 1 kg is available, 1 million casualties will be the result; (2) for the most effective biological agent one particle in the size range 1-5 micrometer inhaled would be effective. [7]. An exercise showed that a bioterrorist attack with smallpox could kill 80 million [8]. On this way, the threat of terrorism with WMD rests on worst-case analysis of every conceivable scenario, and consequently leads to the perception of extreme vulnerability of society.

The focus on the worst-case scenario and corresponding vulnerability perception of a WMD terrorist attack has another important implication: it affects a state response planning. Clearly, governmental policies, organizational, and operational capabilities will vary from country to country. This will color their countering of the threat. Combat against any terrorist threat usually stands on a "total defense" approach, which combines the measures of deterrence, response and restoration into a single integrated program. Such programs are designed to make CB terrorist threats less likely to appear and to create operational capabilities that give a reasonable chance of detecting, defeating and minimizing the consequences. In some cases the program is integrated with routine disaster planning and management.

There are two key prescriptive concepts that have been suggested as guidance for the framing of government response policy: crisis management and consequence management, but not necessarily as separate processes [9].

Crisis management includes efforts to prevent a terrorist attack from occurring; by executing measures to identify, acquire, and plan the use of resources needed; and to anticipate, prevent, and/or resolve a threat or act of terrorism. A strategy for preventing WMD terrorist attacks from occurring is multifaceted: Deterrence involves dissuading states and non-state actors from launching an attack out of fear of forceful international political, economic, and/or military response. Nonproliferation entails using and adapting traditional arms control techniques to stop the spread of CBR weapons, technology, and know-how. Counter-proliferation focuses on more aggressive activities – such as covert action and military strikes – to stop the proliferation of CBR material. Finally, preemption is designed to disrupt an imminent terrorist attack from actually taking place.

Consequence management differs from crisis management in that it involves preparedness and response for dealing with the consequences of a WMD terrorist incident. The main goal of these activities is to protect public and individual health and safety affected by the consequences of WMD terrorism. This is a complex issue because it involves dozens of governmental departments and agencies, fire services, private relief organizations, handling of the press, and a host of other issues. State and local first responders usually include police, firefighters, emergency medical services, and hazardous material technicians. They must identify the agent used so as to rapidly decontaminate victims and apply appropriate medical treatments. If the incident overwhelms local response capabilities, they may call on state or federal agencies to provide assistance.

There is a wide range of issues that impact the effective handling of a CB terrorist incident. The most important of these is that incidents involving infectious agents are likely to be very different from those involving chemical materials. In comparing the effects of chemical and biological agents perhaps the most significant feature is often the immediate nature of the chemical effects whilst the biological attack will only become manifest after an incubation period specific to the material used. Thus both types of incidents would demand different approach in response planning. Terrorist attack using chemical materials is similar in many ways to the hazardous materials incident, and a biological attack is likely to demonstrate the challenges of a major public health emergency.

In the U.S., the threat from chemical/biological terrorism and countermeasures have been given high priority. Although numerous Federal Homeland Security Agencies and Organizations have been involved in combating WMD terrorism, DOD has dominant role in providing technology assistance in WMD defenses. DOD programs for CBR defense are categorized broadly under three operational principles: contamination avoidance, protection and restoration. Detection and warning systems are needed to allow the avoidance of exposure. Individual and collective protection capabilities—including masks, suits, gloves, and collective protection systems and shelters, and medical pre-treatments—are needed to allow forces to operate in contaminated environments. Decontamination systems are needed to prevent the spread or re-aerosolization of agents. Medical diagnostics and treatments are needed to restore force capabilities [10].

In Europe, at present, the threat from WMD terrorism has not been estimated as serious as in the U.S., but this does not mean that Europe can disregard the threat from CBR terrorism but rather that a realistic threat assessment is needed to provide balanced countermeasures. Most European countries still have remarkable CBR defense resources in the framework of military forces, civil defense and emergency services. The protection measures and equipment for civilian populations exist in many countries, first of all, in countries faced with the threat from the former USSR. Performance of this equipment was based on traditional approaches to CBR protection that are guided by military protection philosophy.

Having in mind the changed circumstances of today, there are different needs of civilian organizations for combating WMD terrorism, and higher vulnerability of civilians requires different responses from the military. First of all, it is necessary to develop a rationale and comprehensive civilian risk analysis and more credible chemical exposure scenarios. An integrated approach to the protection system planning is necessary, including CBR technological protection system as well as adequate risk communication and socio-economic aspects of the risk management.

Approach to the WMD terrorist's threat assessment based on the hazard assessment model, often overestimate the hazard, and as a result a robust response capability became necessary. On the other hand this impose the preparedness and response planning in all directions wasting the resources, although there are numerous uncertainties about terrorist threat.

Those uncertainties have been analyzed in a numerous expert studies. GAO summarizes its work and observations in combating terrorism in the report "A Combating Terrorism"[11], as follows: "the attack on September 11, increased the uncertainties regarding the threat. Indeed, there are growing uncertainties regarding the terrorist ability to overcome numerous obstacles to perform large-scale CBR attack and to produce catastrophic effects on the target. Terrorists would have to overcome significant technical and operational challenges to successfully produce and release chemical or biological agents of sufficient quality and quantity to kill or injure large numbers of people without substantial assistance from a foreign government sponsor. In most cases, specialized knowledge is required in the manufacturing process and in improvising an effective delivery device for most chemical and nearly all biological agents that could be used in terrorist attacks".

Based on the given uncertainty about the threat, the GAO report underlines the importance of comprehensive threat and risk assessment as a guide for enhancement of U.S. preparedness against terrorist threats. This approach includes three key elements: a threat assessment, a vulnerability assessment, and a criticality assessment (assessing the importance or significance of a target). Thorough examination of the technical ease of terrorist to conduct an attack with chemical or biological weapons is an essential part of a comprehensive assessment, as well as a better understanding of the motivations and behavioral patterns of terrorists disposed to use unconventional weapons materials. While the risk cannot be eliminated entirely, enhancing protection from known or potential threats can reduce the risk.

3. Principles of Health and Environmental Risk Assessment and Management

The technique of risk assessment is used in a wide range of professions and academic subjects. There is clear distinction between risk and hazard. Hazard is a source of danger that becomes a risk only when there is a finite probability of a manifestation of the hazard. Within this framework, one would then define risk as the product of a hazard and its likelihood of occurrence or probability:

$$\text{Risk} = \text{Hazard} \times \text{Probability.}$$

Risk assessment is the procedure in which the risks posed by inherent hazards involved in processes or situations are estimated either quantitatively or qualitatively. Generally, risk assessment comprises three basic steps: hazard identification, risk estimation and risk evaluation. Hazard identification attempts to identify all the possible outcomes that may eventuate from a particular action. Risk estimation uses analytical methods to estimate the probability of each outcome and the magnitude of the adverse effect associated with that outcome. Risk evaluation is concerned with judging the significance and acceptability of risks, and should include consideration of risk perception and risk benefit studies.

Risk assessment has become a commonly used approach in examining environmental problems. Environmental risk (ERA) is not simply risk to the natural environment. In broad terms, "environment" would be considered as including people and their social and cultural beliefs, as well as the natural environment. The number of hazards that can be examined through ERA is vast, and numerous specific techniques have developed to cope with the characteristics of different hazards. Techniques have also evolved differently due to the institutional basis of the risk assessor and the intended use of the risk assessment. The typology of environmental risk assessments in Europe includes human health risk assessments (HRA), ecological risk assessments (EcoRA), and applied industrial risk assessment (or technological risk assessment-TRA) that examine end-points in people, biota or ecosystems [12].

There are a number of unifying principles underlying all risk assessments presented on Figure 2.

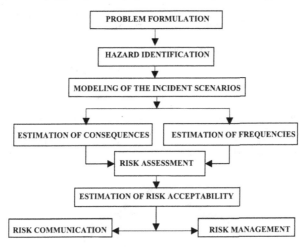

Figure 2 - Steps of environmental risk analysis

The principles of ERA for non-routine incidental releases are to identify the hazard and the release scenario, to analyze the effects or consequences, to provide a quantitative estimation of the event probability and compare it with agreed criteria. Results of an integrated approach to human and environmental risks can be used in risk management decisions intended to protect humans and the environment within defined spatial boundaries.

The evaluation of risk is concerned with issues relating to how those affected by risks perceive them, the value issues underlying the perceived problem, and the trade-off between the perceived risks and benefits. The answers provided by ERA will be crucial in decision-making, and in development of strategies for preventing the incidents and diminishing adverse impacts, if an incident would happen.

Further step in the development of risk assessment methodology has been the comparative risk assessment (CRA), the intention of which has been to identify environmental risks and to rank risks relative to health, the environment, and quality of life. CRA evaluates and ranks environmental risks by using existing information on inherent pollutant hazards, exposure levels, and the population characteristics of an area. The comparative risk process includes a variety of tasks, from problem identification, data collection and analysis, and risk ranking of environmental problems to developing an action plan and implementing new strategies for risk management and reducing risk. Most of the comparative risk projects for priority setting have been initiated by state or local governments and typically by one or more of the environmental protection, natural resource, or health agencies [13].

Risk assessment is carried out to enable a risk management decision to be made. This will depend heavily on how the risk is perceived. Risk perception involves people's beliefs, attitudes, judgments and feelings, as well as the wider social or cultural values that people adopt towards hazards and their benefits. Risk perception will be a major determinant in whether a risk is deemed to be "acceptable" and whether the risk management measures imposed are seen to resolve the problem.

Risk management is the decision-making process through which choices can be made between a ranges of options, which achieve the "required outcome". The required outcome may be specified by legislation, may be determined by a formalized risk-cost-benefit analysis or may be determined by another process. Risk management activities reduce the risk to an "acceptable" level, derived after taking into account a risk perception. Risks can be managed in many ways. They can be eliminated, transferred, retained or reduced.

136

Current efforts to manage risks are often fragmented and sometimes in conflict, often reflecting different statutes. In addition, there is no systematic process for integrating public values, perceptions, ethics, and other cultural considerations into risk-management decisions. It has been argued that the scientific risk assessment process should be separated from the policy risk management process but it is now widely recognized that this is not possible. The two are intimately linked.

The International Standard ISO 14001 (Environmental Management System) represents a new basis for linkage and improvement of environmental risk assessment and management, because it specifies a systematic approach to enable an organization to formulate a policy and objectives for those environmental aspects that can be controlled and over which it can be expected to have an influence [14].

Similarly to this concept, a new model of environmental risk management have been proposed [15]. This model represents a systematic, comprehensive framework that can address various contaminants, media, and sources of exposure, as well as public values, perceptions, and ethics, and that keeps the focus on the risk-management goal.

The proposed framework is intended to be a guide for an approach or thought process for risk management decision-making. It is unlikely that all aspects of the framework would be required for every problem. A number of criteria might be used to determine when applying the framework would be most useful.   The new risk management framework comprises six stages (Figure 3):

Figure 3 - Risk management framework [15]

Different levels of decision-making will require different levels or hierarchy of risk analysis and management, such as: strategic, level, program, and project level. At higher strategic and policy levels "problems" would be defined in general terms, and their decisions have long term implications and consequences associated with considerable uncertainty. At lower program and project levels risks would be better understood, and estimates of risks and corresponding response actions would be based on high quality consistent data for well-defined activities at specific sites. In principle, CRA and ERA are best suited for assessing the risks at higher levels, while HRA and TRA would be appropriate for lower levels. At the same time, information received from impacts noted at the lower levels is fed back into the decision-making process at the higher levels.

4. Application of Environmental Risk Management Principles for Combating Terrorism

Described approaches to the environmental risk assessment and management represent a broad framework which could be used for the development a set of models for assessing and managing the risks associated with WMD terrorist threat. Although the number of CBR hazards and potential actors of terrorist attack that can be examined through risk assessment are vast, in all types of risk assessment there is the basic scheme required in all cases. This scheme for WMD terrorist incident risk assessment and management represents a complex control system, which is consisted of the model of object, risk assessment and risk management model (Figure 4).

The main feature of this framework is the site or object oriented approach, where object represents potential target area with its complex structures and functions. The people in a target area should be divided in categories or relatively homogenous exposure zones that should be characterized according following criteria:

- Similarity of mission during WMD incident (first responders, general public, and others),
- Similarity with respect to hazardous agents and exposure level,
- Characteristics of environment (open spaces, public buildings, transportation means and objects, etc.),
- Population characteristics (age, health status, sex).

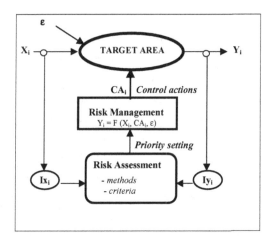

Figure 4 - Basic scheme for WMD risk assessment and management

The proposed framework first requires that a potential or current problem be put into a broader context of public health or environmental health by determination of the set objectives and goals of control. In order to define the structure of the object, it is necessary to recognize which inputs and outputs of the object are going to be included into its model. The object of control has three inputs: the observed inputs (X_i), observed and controlled inputs (CA_i), and unobserved inputs ε. Outputs represent the state of the object, that are dependent on its inputs, both controlled and uncontrolled: $Y_i = F (X_i, CA_i, \varepsilon)$. The information on the state of object's inputs (Ixi) and outputs (Iyi) represent indicators that would be used to describe the magnitude of the problem, provide the basis for setting objectives and goals, as well as inputs in risk assessment model.

Unlike hazard assessment methodology which formulates indicators that address the fundamental causes and effects of a CBR agent, risk assessment use a probabilistic approach that defines various indicators to identify which specific attacks are most likely and what severe consequences could take place. There are many considerations that may take into account when selecting indicators for evaluation and management WMD incident risks. In the context of combating WMD terrorism indicators would be selected from three types: (1) health effects indicators for estimation of health risks, (2) operational performance indicators that provide information on the performance of technological protective systems, and (3) management performance indicators for evaluation the performance of implementation the policies and programs for combating WMD terrorism.

For example, for estimation of health risks from certain CW terrorist attack or hazardous material incident as health effects indicator can be used:

- Dose-effect function, which is a way to estimate the relationship between the amount of an agent received in an exposure (dose) and nature or severity of the resultant harm (effect), or

- Dose-response function, which is the relationship between dose (exposure) and the percentage of the exposed group that shows a pre-defined effect from that dose.

Common practice in the managing hazardous materials incidents is still based on hazard analysis and consequence assessment using as health risk indicators total number of people affected (population at risk), number of injured and/or fatalities based on the dose-response relationship [16, 17, 18].

Sophisticated methodologies have been developed, many of them involving mathematical and statistical modeling on computers, to support the risk assessment. As an example, one of mathematical models is presented for calculation of collective risk $R_k(x,y)$ by following equations [19]:

$$R_k(x,y) = \Sigma\ R_i(x,y)\ \cdot\ N(x,y) \qquad (1)$$
$$R_i(x,y) = \Sigma\ \lambda_i\ E_{ij}(x,y)\ \cdot\ F_j \qquad (2)$$

where are:
$R_i(x,y)$ = individual health risk at site defined by coordinates (x,y)
$N(x,y)$ = number of people per unit surface at site (x,y)
λ_i = probability of event according scenario "i",
$E_{ij}(x,y)$ = probability of impact "j" at site (x,y) during realization of scenario "i",
F_j = probability of fatal consequences during realization of impact "j".

The equation (1) shows that important factors for collective risk calculation are population density at a site and a predicted scenario of the WMD incident. Such approach enables preparation of a map of the homogenous fields of risk, where "field of risk" represents distribution of individual risks on defined target area.

On the other hand, risk assessments inevitably incorporate uncertainties, often extremely large, in their estimates of exposure and risk. Most of these uncertainties arise because of unobserved and uncontrollable inputs to the system under consideration (ε in Figure 4). Scientific measurements needed to determine risk precisely cannot be made in the actual individuals, exposure setting, and time frame of interest. As alternatives, expert methods (checklist, matrices, threshold analysis, etc.) are used, from which answers must be extrapolated to the setting of interest. The magnitude of the uncertainties from such extrapolations is typically unknown. Thus, when any risk assessment is complete, no one knows how close its estimates are to the exposures and risks present in the real world.

The most common have been used ranking methods based on qualitative measures of the severity of the consequences and the likelihood of the WMD terrorist attack, as exemplified in Table 1.

Table 1 - Example of the classification of consequences and frequency of the chemical terrorist incidents

Consequences	Description	Class	Frequency	Description	Class
Insignificant	No effects	I	Ounce a year	Almost certain	A
Minor	Low level short term disability, no medical treatment required	II	Ounce in 1-10 years	Likely	B
Moderate	Minor disability requiring medical attention	III	Ounce in 10-100 years	Possible	C
Major	Extensive number of revertible injuries or several fatalities	IV	Ounce in 100-1000 years	Unlikely	D
Extreme	Extensive number of irreversible injuries or multiple fatalities	V	Less than 1 in1000 years	Rare	E

A range of risk levels can be identified by the combination of consequences and frequencies, and usually risks have been ranked, as follows:

1) Extreme/significant risks,
2) High/medium risks, and
3) Low or insignificant risks.

The final part of risk assessment and management scheme is a risk management model, which is concerned with: "what we can do about risk", i.e., finding ways to achieve the desired level of "safety". Risk management attempts to analyze which options for action based on the results of the risk assessment will produce these pre-determined risk levels or simply learn to live with risks.

Risk-management options and management objectives are interconnected, and setting management objectives without understanding if or how they can be achieved may result in unrealistic objectives.

Contemporary U.S. strategy for combating WMD terrorism is still based on the "zero risk" concept that requires "total defense" approach, which concentrates merely on elements of prevention an attack from occurring (e.g., arms control), or on domestic response preparedness (e.g., training emergency responders, or managing the consequences of an attack). Congress authorizes (2001) the agencies to provide technical assistance and monetary grants for emergency planning, training, and equipment acquisition [20]. Federal preparedness programs fall into the following categories:

- Emergency management and planning,
- Training and equipment for first responders,
- Weapons of mass destruction and hazardous materials,
- Law enforcement, and
- Public health and medical community.

Programs for chemical and biological defense have the following key technology initiatives for combating chemical and biological terrorism: contamination avoidance, protection, and decontamination equipment packages; emergency response capability for consequence management, and others. It is interesting that these programs did not include any personal protection equipment or medical treatment first aid devices for civilian populations.

5. Requirements for the Development of Personal Protective Equipment for Civilians

There are four key cornerstones of the technology for personal and collective protection against chemical, biological and radiological agents and weapons:

- Physical protection: respiratory protection, body protection, sheltering;
- Medical protection: pretreatment, therapy;
- Detection: alarm, monitoring, verification, identification;
- Decontamination and disinfections.

In contrast to the military and first responders, a process to develop clearly defined requirements for the technological protection systems of civilian populations against WMD attack in U.S. does not yet exist. In some European countries traditional approach to the CBR protection of civilian populations in the framework of civil defense is still present. For example, during the Cold War most European governments faced with Soviet WMD threat made the decision that the whole population in the state should have access to individual protective equipment. This equipment was designed according to high military requirements, based on the worse-case approach that included: extreme toxicity of agents, maximum exposure levels, lowest threshold exposure values of agents, long duration of exposure, requirement for high level of protection, etc. The resultant protective equipment, of course, was very similar to military equipment at that time, and complicated for use by civilians. According to a Swedish study, only 38% of untrained participants knew how to use a protective mask [21]. Besides, many of the inhabitants of Tel Aviv-Israel did not know how to use protective masks during the Iraqi bombing in the Gulf Crisis 1991, and some of them had health consequences, although they were not actually exposed to any toxic agent.

Similar to the respiratory protection equipment, all other measures and devices of the CBR protective system have had performance criteria based on the worst-case scenarios, and each protective measure and device has been developed to possess the highest performance as possible. As result, the overall CBR protective system effectiveness could not be satisfactory in many cases, despite the good performance of particular equipment. There are various reasons, but two are most important: (1) many of the vulnerable groups can not be supplied with any kind of protective equipment nor medical treatment devices, and (2) the European experience has shown that without proper training, the efficiency of equipment usage can not be satisfactory.

A proposed new approach to the development of the criteria for the development of integrated technology system for combating CBR terrorism is based on the previously described framework for WMD risk assessment and management (Figure 2 and 4). Although the consequence management of a terrorist incident should cover all known and potential chemical, biological and radiological weapons and agents, main idea will be explained specifically for the case of chemical attack.

It is reasonable to suppose that first responders would be at extreme risk (level 1), while various groups of general population would be distributed through lower risk classes (levels 2 and 3). In principle, the general goal of protection today is to prevent poisoning of all potentially exposed people. But, following the classification of risks in various levels, it is possible to offer various management goals. The criteria for measurement of success of the risk, management according health effects indicators, could be different for these three risk levels. For example, for first responders it is essential to ensure performance of the mission in the contaminated area. Reduction of health risks of first responders should be complete, and the residual risk has to be zero. That corresponds to the NOEL (non-observable exposure level) represented in Figure 5. In that case control action should be use personal protective equipment that has the highest performances, equivalent to the PPE level A commonly used by US HAZMAT responders.

On the other hand, management goals for the protection of the general population at risks in levels 2 and 3 "to protect human health and/or save life" may be satisfactory. The criteria for the estimation of health effects indicators could be different, although the OSHA Short-Term Exposure Limit (STEL) and the NIOSH Immediately Dangerous to Life and Health (IDLH) levels have been commonly used worldwide as exposure limits in the case of industrial chemical accidents.

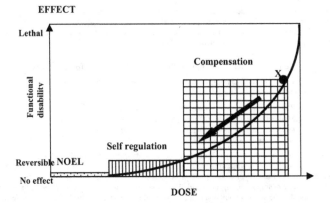

Figure 5 - Dose to effect relationship

However, these exposure limit values were developed for an acute occupational exposure and there are certain limitations in its application to non-occupational situations. From this reason, several other exposure guidelines levels have been developed which are applicable to the general public, including children and the elderly. Some examples are: AIHA - Emergency Response Planning Guidelines (ERPG); NAS-Short Term Public emergency guidance Level (SPEGL); NAS-Military Emergency Guidance Level (EEGL); TTCP- Temporary |Tolerable Concentration of Pollutant and others [16, 22, 23].

The AIHA defined three ERPG levels that vary with the health effects expected with exposure could be correlated with designated fields on the dose-effect diagram (Figure 5) as follows: ERPG-1: transient symptoms/self-regulation; ERPG-2: ability impairment/compensation; and ERPG-3: life-threatening/ point X or higher doses. These exposure levels can be also compared with estimated incidents consequence levels III, IV and V (Table 1), and could be used for defining the health effects indicators (dose) and basic operational performance indicators of technological protective system (protection factors).

The protection factor of overall CBR protective system (PF_{tot}) can be defined as a measure of prescribed (nominal PF) or realized (field PF) reduction of exposure, dose or health risk. Nominal PF_{tot} is dependent on efficiency or protection factors of particular protective measures as follows: detection (PF_{det}), respiratory protection (PF_{rp}), body protection (PF_{bp}), collective protection (PF_{cp}), decontamination (PF_{decon}), and medical treatment (PF_{medt}). In practice each of these PF's have defined as relationship between assault dose (or airborne exposure) and standardized exposure limit values for particular substances (PFi> airborne dose/ threshold dose). Such requirements for protection efficiency led to independent development of each subsystem, which are too sophisticated to be used in the chemical incident for protection of general population.

The new approach recognizes different needs for protection of general populations in different exposure zones, and on that basis is possible to define requirements for performances of particular protection measure by lower particular PFi so that overall protection factor PF_{tot} satisfy exposure standards and realize pre-determined goals of risk management. Complex nature of chemical protective system and numerous posibilities of serial-paralel linkages between particular protective measures demand more serious analysis.

As an example, a determination of protection factor for respiratory protection could be determined at the various levels in dependence of the type of exposure zone or risk level, as shown in Table 2.

Table 2 - Relationship between risk level and protection factors for respiratory protection

Risk level	Assult exposure level	Exposure level limit	Protection factor	Additional treatment
Extreme/significant risks	High	ERPG-1	High	No treatment
		ERPG-2	Medium	Medical treatment
		ERPG-3	Low	Medical treatment
High/medium risks	Medium	ERPG-1	Medium	No treatment
		ERPG-2	Low	Medical treatment
		ERPG-3	RPE- N.A.*.	Medical treatment
Low or insignificant risks	Low	ERPG-1	Low	No treatment
		ERPG-2	RPE- N.A..	Medical treatment
		ERPG-3	RPE- N.A..	Medical treatment

* Respiratory equipment not available

Estimates shown in the table allow wide possibilities for selection of respiratory protective devices (military as well industrial) with different protective performance [24, 25]. But, protective performance is not only criteria that should be used in the protective device selection. There are also other characteristics of respiratory protective equipment that are important as criteria for respiratory protection planning and respirator selection: respiration resistance, mask dead space, comfort , vision, speech transmision, use of corrective lenses, liquid drinking , handling, mass and dimension, durability, reliability, and price [26].

6. Conclusion

Considerable energy has been applied to counter the danger of terrorists using of unconventional weapons materials. However, considerably less effort seems to have been made to fully understand the nature of the WMD threat. The reason is that much effort to improve our capabilities to combat the threat of CBRW terrorism, and to manage the consequences should an attack ever occur, has been based on worst-case scenarios. These worst-case scenarios tend to be shaped by perceptions of vulnerabilities and the availability of technology for response to a CBW attack and do not take into account other key factors. As result, such approach usually overestimated the significance of the source and concentration of contaminant, as well as potential for a contaminant exposure pathway and vulnerability of potentially affected groups.

This paper analyzed key characteristics of hazard assessment and risk assessment and management methodologies. There is proposed a framework for WMD terrorist incident risk assessment and management based on environmental

142

risk management principles. The essential characteristic of existing CBR technologies were also discussed, and critical review for their application for the protection of civilian populations in WMD incident was done. There is proposed new approach for defining basic requirements for protective measures, based on dose-effect relationship, and on set of protection factors.

New approaches to risk assessment and management will be new challenges for science and technology to define basic requirements and solutions for the development of an integrated management and technological system for combating WMD terrorism, as follows:

- to establish an information system that will contribute to better understanding of hazards and risks from WMD terrorism,
- to develop a specific methods for WMD terrorism risk assessment, including new standards for short term exposure limit levels,
- to evaluate and rank population groups at risk,
- to establish zoning system and mapping based on the exposure assessment,
- to develop and standardize different types of PPE for various risk zones,
- to develop new approaches for improvements of detection, decontamination and medical treatment.

The results of these activities should improve the existing strategies for combating WMD terrorism accommodated to more reliable risk assessment as cornerstone of whole system. Information from risk assessments should: contribute to convicting terrorists that they will not achieve their objectives, rise up the motivation of the general public to take part in preparedness for their own protection and safety, help to the government and local authorities to make more effective response plans, improve protective equipment and methods, improve local training programs for combating WMD terrorism, as well as response capabilities of emergency responders and local communities.

7. References

1. Antony Tu, Procedings of the 6th Int. Symp. Prot.Chem. Biol. Warfare Agents, Stockholm, Sweden, 15-19 May, 1998

2. Staten L.C., Emergency Response to Chemical/Biological Terrorist Incidents, http://www. ERRI "Lessons on Line" - # 1-97, 197

3. Planin E., Washington Post, Tuesday, March 12 2002, Page A08

4. Parachini J.V., Combating Terrorism: Assessing the Threat, Center for Nonproliferation Studies, Monterey Institute of International Studies, (CNS Report), October 20, 1999

5. Financial Times, August 27, 1999

6. Roffey R., Bioterrorism a Swedish view, is the threat overstated?, The 7th Int. Symp. Prot.Chem. Biol. Warfare Agents, Stockholm, Sweden, 15-19 June 2001, Procedings CD

7. Medema J., Minimum Military significant amount of agent, threat reduction through physical protection, CWC and BTIFFC, Proc. 7th Int. Symp. Prot.Chem. Biol. Warfare Agents, Stockholm, Sweden, 15-19 June, 2001, Proc. CD

8. Bioterrorism exercise showed smallpox could kill 80 million, San Antonio-Express News, February 22 (1999)

9. Cilluffo F.J., S.L.CardasH, G. N. Lederman, Combating Chemical, Biological, Radiological and Nuclear Terrorism: A Comprehensive Strategy, Center for Strategic and International Studies, Washington D.C., 2000

10. Anna Johnson-Winegar (2001), 7th Int. Symp. Prot.Chem. Biol. Warfare Agents, Stockholm, Sweden, 15-19 June, Proc. CD

11. Hinton H.L., Combating terrorism: Considerations for Investing Resources in Chemical and Biological Preparedness, Testimony Before the Committee on Governmental Affairs, U.S. Senate, GAO, October 17, 2001

12. Fairman R., Mead .C., W.P. Williams, Environmental Risk Assessment: Approaches, Experiences and Information Sources, European Environmental Agency, Copenhagen, 1998

13. Environmental Protection Agency (U.S. EPA), Unfinished Business: A Comparative Assessment of Environmental Problems, U.S. EPA 1987b

14. ISO 14001:1996, Environmental management systems - Specification with guidance for use

15. Carnegie Commission on Science, Technology, and Government, Commission on Risk Assessment and Risk Management, Risk Assessment and Risk Management in Regulatory Decision-Making, New York, NY, June 13, 1996

16. U.S. Department Of Health And Human Services, Public Health Service, Agency for Toxic Substances and Disease Registry, MANAGING HAZARDOUS MATERIALS INCIDENTS, Volume I, March 2001

17. Regulation on methodology for hazard assessment from chemical accident and pollution of the environment, Official Gazzette of Republic Serbia, No. 69, 1994

18. Environment Health and Safety Manual, Section 3. Identification, Assessment and Control, ABN: 84 002 705 224, The University of Melbourne, Australia, 2000

19. V.S.Safronov et al., Theory and practice of risk analysis in gas industry, NUMC Minprirody, Moskva, 1996

20. CRS Report for Congress-Terrorism Preparedness: A Catalog of Federal Assistance Programs, December 27, 2001

21. Rejnus L., Thorpsten J., Abrahamsson A., Individual Preparedness for Protection Against Toxic Agents: Summary of the study, Proc. 7th Int. Symp. Prot.Chem. Biol. Warfare Agents, Stockholm, Sweden, 15-19 June, Proc. CD, 2001

22. Stojanovic B., Problems of using the pollutant limit values for estimating the hazard in chemical accidents, Arch.Toxicol.Kinet.Xenobiot.Metab., 2(2) pp 91-92, 1994

23. Decker J., H.Rogers, application of revised airborne exposure limits for chemical warfare agents, Book of Abstracts-The Fourth International Chemical, Biological Medical Treatment Symposium, 28 April-3 May 2002, Spiez, Switzerland

24. M.Jovasevic-Stojanovic, B.Stojanovic, Personal Protection Program against Toxic Industrial Chemicals in the Case of Chemical Accidents, Book of Abstracts-The Fourth International Chemical, Biological Medical Treatment Symposium, 28 April-3 May 2002, Spiez, Switzerland

25. Stojanovic B., M.Jovasevic-Stojanovic, Review of the Development of Chemical Weapons and Personal Protective Equipment and Future Perspectives, Book of Abstracts-The Fourth International Chemical, Biological Medical Treatment Symposium, 28 April-3 May 2002, Spiez, Switzerland

26. Stojanovic B., M.Jovasevic, Lj.Djuricic, M.Polovina, How Well Do Respiratory Protective Devices Really Work ? International Conference on Emergency Civilian Medical Services in a Non- Conventional War, Tel Aviv, Israel, 1992, Program and Abstracts,p.20

COMPONENTS OF PREPAREDNESS –
IMPORTANCE OF DISEASE REPORTING AND EPIDEMIOLOGY CAPACITY

FILIZ HINCAL
University of Hacettepe, Faculty of Pharmacy, Department of Toxicology,
Hacettepe Drug and Poison Information Center
ANKARA – TURKEY

Abstract

After the 9/11 disaster, not only the US, but also all the nations and all the people of the world realized that the threat of terrorism; chemical, biological, radiological or nuclear (CBRN), is real. With the understanding of the importance of preparing the public health infrastructure to prevent illness and injury, especially from a covered biological terrorism, a much vigorous effort is being made to improve and re-examine core public activities, information and detection systems in all developed countries. Epidemiology reporting system is one of the most effective means of combating CBRN terrorism, particularly with biological type. In this paper, the necessity and requirements for establishing and running of effective disease reporting and epidemiologic investigation systems in developing countries, and the major problems faced will be discussed.

1. Introduction

The main lesson of the September 11 disaster is that there is no nation or state that is secure or immune from any kind of terrorist action. Not only the USA, but all the nations of the world are vulnerable to terrorist attacks, which could entail the use of chemical, biological, radiological, or nuclear (CBRN) weapons. Today, those weapons of mass destruction (WMD) are readily available to many countries, and WMD are a viable alternative to conventional weapons for most terrorist groups. As long as the conflicts, ignorance, poverty, socio-economical imbalances and political instabilities exist, this seem to continue. Thus, the threat is greater than ever today, with potentially devastating consequences, including enormous number of deaths and widespread diseases, and destruction of public health infrastructure.

Although there is no absolute protection against all possible CBRN agents, "preparedness" is essential to protect and reduce the harm in case of an attack, and the lessons of the past have thought everybody that preparedness by accumulating knowledge, equipment and an efficient "system" is mandatory to combat WMD and terrorism of all kinds.Biological terrorism differs from any other types of CBRN terrorism in that it would impose particularly heavy demand on the nations' public healthcare system, and impose stressful burdens.

Understanding and quantifying the impact of a bioterrorist attack are essential in developing public health preparedness. With a covert biological agent attack, the most likely first indicator of an event would be an increased number of patients presenting with clinical symptoms caused by the disseminated disease agent. Therefore, health care providers must use epidemiology to detect and respond rapidly to a biological agent attack; understanding of the basic epidemiologic principles of biological agents used as weapons is critical to effectively counter the potentially devastating effects.

Many diseases caused by weaponized biological agents present with nonspecific clinical features that could be difficult to diagnose and recognize as a biological attack. The disease pattern that develops is an important factor in differentiating between a natural and a terrorist or warfare attack. Although the recognition of and preparation for a biological attack is similar to that for any disease outbreak, the surveillance, response, and other demands on resources would likely be of an unparalleled intensity. A strong public health infrastructure with epidemiologic investigation capability, practical training programs, and preparedness plans are essential to prevent and control disease outbreaks, whether they are naturally occurring or otherwise (1).

P. Stopa and Z. Orahovec (eds.), Technology for Combating WMD Terrorism, 145-148.

2. Components of a Health Response System

Major components of an ideal public health and medical response system may include surveillance, epidemiology, laboratory capability, medical management, training/education and information/communication (2).

2.1 SURVEILLANCE

Earlier detection enables earlier recognition of the nature of the incident by epidemiologists and laboratory personnel and in turn, enables a more effective and efficient response. In order to provide this early warning function, a health surveillance system that enables the development of data baselines to establish normal health status, detection of minor changes, and continuous monitoring of health of the population need to be established. Such a system may collect information (unusual cases of illness, number of admissions or visits, animal illness incidents, sick calls, purchases of medications, infectious disease related unusual deaths, etc) from various resources and providers (hospitals, emergency departments, physicians and veterinary offices, pharmacies, schools and employers, etc), then, establishes systems of exchanging and analyzing this information.

2.2 EPIDEMIOLOGY

Whereas health surveillance systems provide the awareness or a detection tool, epidemiology serves as an assessment tool used to ascertain the exact nature of a bioterrorist event. Epidemiology/ epidemiologist interprets raw data gathered through surveillance and investigations, and recognizes the importance of the event, determines the source of the outbreak, mode of transmission, extent of exposure and pattern of progress. Based on this information, recommendations are made for the appropriate public health measures needed to contain the outbreak, and treatment.

2.3 LABORATORY CAPACITY

Laboratory is another essential tool of the response system to bioterrorism. Physician's decision and treatment will be based on the detection of agents involved. Capability of testing for antimicrobial sensitivity and determining the effectiveness of available antibiotics and vaccines is also necessary. To organize a laboratory response network, which consists of a series of laboratories of varying capabilities and assisting each other through cooperative rearrangement, is essential for an efficient system of response. Since rapid diagnosis is critical, the existence of adequate number and quality of multilevel laboratories is highly important, whose capabilities should be continuously upgraded and continuous training of technicians should be provided.

2.4 MEDICAL MANAGEMENT

Medical response to bioterrorist actions involves triage, prophylaxis, treatment and logistics components, all of which are interrelated and dependent each other. Providing timely and adequate medical management is a challenging task and requires vigorous preparedness with knowledge, equipment and planning and organization.

2.5 TRAINING AND EDUCATION

Early recognition of biological attack depends on two critical resources: epidemiological warning system and individual clinical expertise of medical personnel. Therefore, education and training of medical personnel of all categories on the recognition of agents and outbreak, treatment of casualties and self-protection as well as the organization of the whole response system to enable them acting most efficiently is mandatory.

3. Epidemiological Investigation

If we focus on epidemilogy further: once a biological attack or any outbreak of disease is suspected, the epidemiologic investigation should begin immediately. The first step is to confirm that a disease outbreak has occurred. A case definition should be constructed to determine the number of cases and the attack rate. The estimated rate of illness should be compared with rates during previous years to determine if the rate constitutes a deviation from the norm. Once the attack rate has been determined, the outbreak can be described by time, place, and person. The epidemic curve is calculated based on cases over time; the early parts of the epidemic curve will tend to be compressed compared with propagated outbreaks. The peak may be in a matter of days or even hours. Later phases of the curve may also help determine if the disease appears to spread from person to person, which can be extremely important for determining effective disease control measures (3).

The performance of epidemiology is highly dependent on the quality of the surveillance data and   how quickly it is received is critically important. In order to function properly, capabilities of interpreting and continuous monitoring of surveillance data, conducting investigations and building case definitions are needed. Epidemiology has a central role in determining the point of initial exposure and localizations where to focus prophylaxis and treatment first, measures for containing the outbreak, clues for a low enforcement investigation, etc. Following a suspicion of an attack, a case definition is made to alert public health authorities and medical personnel, and treatment protocols and advanced clinical symptoms are provided. Epidemiologists' service continues thereafter, to provide information back to public health and medical personnel; however, this needs continuous access to information of patients' loads and symptoms and laboratory results. In other words, epidemiology is a continuous liaison between surveillance data and response and treatment teams that provides the most appropriate assessment for the basis of correct and prompt decisions. In order to achieve all these goals, the epidemiology service or the component of the response system ideally needs:

- Real-time access to surveillance data

- Adequate personnel to analyze surveillance data and investigate unusual outbreaks,

- Broad knowledge of a variety of disease patterns

- Electronic systems to compile and analyze patient data

- Information exchange and communication between laboratories, hospitals, physicians, public health services

Therefore, to establish and improve epidemiological capabilities of a nation against bioterrorism, the following basic initiatives should be undertaken:

- Improvement of disease reporting systems

- Funding of local and central public health departments to hire and maintain adequate number of qualified personnel

- Identification of epidemiological thresholds for triggering particular responses

4. What are the Major Issues in Establishing Such a System?

Implementation of such an epidemiology system is a great challenge in many respects, and it is not an easy task at all, particularly for the developing countries. Although, numerous measures to improve preparedness for and response to biological warfare or terrorism are ongoing everywhere, and training efforts have increased both in the military and civilian sectors, it is known that, at present, even in the well-developed countries, very few localities have established epidemiology services, electronic systems for epidemiology, and public health and medical entities to receive or exchange surveillance data. Well before any event, public health authorities must implement surveillance systems so they can recognize patterns of nonspecific syndromes that could indicate the early manifestations of a biological warfare attack. The system must be timely, sensitive, specific, and practical. To recognize any unusual changes in disease occurrence, surveillance of background disease activity should be ongoing, and any variation should be followed up promptly with a directed examination of the facts regarding the change. Major problems faced in developing countries are as follows:

- Adequate personnel with proper qualification is in short supply

- Surveillance data collection needs a comprehensive organization system with personnel and equipment and a central body, which will work timely and efficiently, as well as individual data collections in medical units, hospitals, etc.

- Establishment of a comprehensive, efficient and friendly electronic system is not easy and inexpensive but, once the organization frame is funded, it may not be so difficult to implement and maintain a very basic but efficient model

- Another important problem is the number and quality of laboratories and a laboratory response network that the surveillance data and epidemiological assessment will be based on. Equipment and expertise is the most cumbersome aspects of this system to implement, particularly with its high cost

5. Conclusion

In conclusion, to build and run an efficient system of disease reporting and epidemiological assessment requires a well-organized, self-sufficient health care system. Without solving/improving the infrastructure of health care provisions in a given country, all efforts would be limited and with limited success, and the incidents will be highly effective and casualties and losses of all kinds would be highly and costly. However, we should always remember that the global biological warfare threat is serious, and globalization of its effects is inevitable, and the cost will be paid much or less by everybody.

6. References

1. CDC (2000) Biological and Chemical Terrorism. Strategic Plan for Preparedness and Response. MMWR 49, RR-4.

2. Chemicals and Biological Arms Control Institute Final Report (2000). Bioterrorism in the United States. Threat, Preparedness, and Response.

3. U.S. Army Medical Research Institute of Infectious Diseases (2001). USAMRIID'S Medical Management of Biological Casualties Handbook. 4th Ed. Fort Detrick, Frederick, Maryland

AUTOMATED DECISION AID SYSTEM FOR HAZARDOUS INCIDENTS (ADASHI):
An Incident Response and Training Tool for Chemical/Biological Hazards

JAMES GENOVESE
Edgewood Chemical Biological Center
Aberdeen Proving Ground, MD

ARTHUR STUEMPFLE,
OptiMetrics, Inc.
Abingdon, MD

1. Introduction

Defense Secretary William Cohen has long expressed concern about the threat posed by weapons of mass destruction and the country's vulnerability to an attack on its own soil. In 1997, Cohen commented that the threat posed by the proliferation of such weapons, is "the greatest threat that any of us will face in the coming years". A year earlier, Congress passed the Defense Against Weapons of Mass Destruction Act, which designated the Department of Defense as lead agency for responding to an attack by terrorists using chemical or biological hazards. As a result of the law, known as the Nunn-Lugar-Domenici Act, the Defense Department created the Domestic Preparedness Program to train local and state officials most likely to be first responders in the event of such an attack.

Emergency responders and incident command personnel must make rapid decisions in life-saving and life-threatening situations. When dealing with terrorist incidents involving suspected or known chemical-biological hazards, these personnel require training in the use of and rapid access to technical knowledge to properly react to these types of hazardous events.

The integration of disparate functional responders such as medical and decontamination specialists, hazardous material (HAZMAT) teams, and explosive ordnance disposal (EOD) teams requires substantial planning, coordination and practice. This dependency and interdependency of multiple operational functions are mapped out in Figure 1. The need exists to integrate hazard projections and challenges to enable the production of requirements.

Current reference documentation and formal training curricula are singularly structured to the point of being "stove-piped". For example, decontamination processes and physical protection requirements are not normally linked but are taught as independent functions.

In practice, however these processes are integrally linked. The need exists for a logic base that tactically relates these functional elements together based on rapidly collected and identified information as responders assess the situation.

The Automated Decision Aid System for Hazardous Incidents (ADASHI) is a portable, computer-based decision support/training system for improving the response to a hazardous incident by military and civil responders. ADASHI is designed to function on laptops and desktop computers and can be used at the site by the incident commander (IC) or at higher echelon operation centers. The tool has the capability to support individual and collective training at detachment or team locations or even at a responder's home. This system integrates the specific technical functions required to manage a hazardous incident involving chemical or biological hazards. Those functions include:

- Initial hazard assessment
- Hazard source analysis
- Physical protection
- Decontamination methods
- Hazard area prediction

P. Stopa and Z. Orahovec (eds.), Technology for Combating WMD Terrorism, 149-156.

- Detection planning and sampling
- Medical treatment, and casualty estimation criteria
- Hazard Mitigation.

Specific functional inputs are integrated with decision criteria. The initial development has been on Chemical incidents. Approximately 50% of the required functionality has been established. Other Weapons of Mass Destruction (WMD) categories, such as response to biological agents are in the formative stages of development.

ADASHI is designed to automatically monitor the essential aspects of an event, whether it be a "What if" simulated event for training purposes or a real event. This automated time-critical and time-dependent control and monitoring function is accomplished using detailed computer algorithms and data processing architecture requiring embedded expert assistance logic and multivalent neural processing techniques. These neurofuzzy algorithms act on immediate "human" inputs (i.e. imperfect, less than optimal) linguistic variables. Automated multifunction tracking and monitoring is valuable as a training tool where individual data inputs can influence a WMD training scenario outcome. The system can then help determine the scope of operational alternatives available and query the trainee using direct questions, memory prompts, etc. to help in making an informed decision. The trainee must then select his actions from the options in order to mitigate the effects of the incident. The database structure alleviates the training burden by offering in electronic format disparate reference material. Team leaders and members can perform "trial and error" learning and build confidence and expertise in different learning environments.

ADASHI is to be utilized as an "over the shoulder" decision-support system to aid incident commanders by processing the multivariate input data and providing critical information to the system user in a high-stress environment. This system can also be employed at higher echelons, such as operations centers or 911 dispatch centers, to aid in community response resource management.

In the remainder of this paper, general design features will be discussed then the training overlay will be presented. Finally, an example of a training scenario will be worked through.

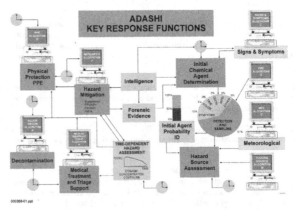

Figure 1. **Key response functions as an integrated framework.**

2. Design Concept

The graphical user interface (GUI) enables the user/trainee to identify the operational situation by entering information peculiar to their circumstances. There are two types of data that are processed – high confidence data such as time, meteorological conditions, and physicochemical characteristics, and other softer/fuzzier data originating from a variety of human sources such as 911, other initial responders or victims at the scene. The user interface exists solely to collect and present information with all algorithms implemented as separate objects. The intent is to use as much non-numerical or linguistic variable information input as possible. The logic here is that inputs during the chaotic notification and response phases of an incident will be fuzzy anyway. Therefore, it makes sense to process this less than optimal information to "get in the ballpark", as far as how an incident is progressing.

The user will have direct access to memory aids and check-off lists or can make queries of the libraries contained within this architecture. The user will make the choice up-front as to whether to use the tool as a training instrument or in the decision mode. In particular, if the tool is used in training, the software is configured to ask questions of the trainee. For example: what response to take and then compare that to "the suggested school solution". When the tool is applied in the decision mode, the suggested responses are provided as output from the behind-the-scenes computations. As an example in the proof of principle version, the incident responder is presented with a checklist of potential symptoms   experienced by casualties as a matter of time from agent exposure and observable material properties (signs). The processing software then "determines" the probable causative agent based on the symptoms and related "signs" selected from the checklist, or indicates an unknown agent.

Appropriate physical properties are then pulled-up from the agent characteristic library and used in computing time-dependent vapor concentrations following agent dissemination in a ventilated enclosure. In the training instrument mode, the user could be asked to identify the likely agent based on the selected set of symptoms defined in the established training scenario or vignette.

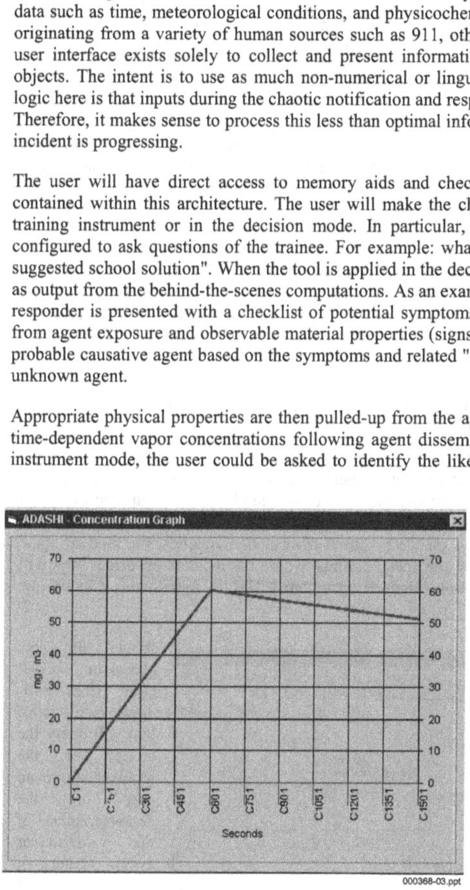

Figure 2. Time-based dosage assessment.

Existing hazard prediction models will be used where possible. For outdoor incidents, the user can choose from either the Personal Computer Program for Chemical Hazard Prediction (D2PC); the Vapor, Liquid, and Solid Tracking (VLSTRACK) or the Hazard Prediction Assessment Capability (HPAC). **However, it is predictable that many terrorist incidents will occur below the threshold of these models.** Prediction tools that function within urban sectors or neighborhoods will be included. Those readily available will be evaluated for use. Currently, an enclosure model InDeVap [Indoor Evaporation Model] is available for indoor release of hazardous chemicals. Other PC-based indoor models will be evaluated for applicability. A multiple domain hierarchy is required for hazard prediction models and terrain databases. The Hazard Source Assessment processor will integrate a variety of hazard prediction models such as HPAC, D2PC, VLSTRACK, InDeVap, etc. As a result, a time-based assessment of environmental hazard dosage (Figure 2.) can be presented. ADASHI is configured such that similar or identical input icons will drive the selected prediction model. Outputs can be presented as an integrated time/concentration contour plot.

2.1. AGENT SIGNS AND SYMPTOMS

The initial signs and symptoms are very important in order to focus on the true nature of the hazard and eventually its identification. In many cases, a combination of signs and symptoms are so specific, that it suggests what has happened and what detectors will best verify the presence of a particular hazard. In the initial prototype, a time-based listing of symptoms for exposure are used to determine the chemical hazard. A drop-down window listing the **symptoms** allows the user to "check off" (indicate) prevalent symptoms from incident casualties and observable signs of the hazardous material. This module tracks both the exposure time, onset time to effects and symptoms; and uses the signs and symptoms combination to make an estimate of the type of hazard involved in the incident (i.e., nerve agent, tear gas, etc.).

Through the Expert Assistance algorithm and database comparisons, a more refined determination of the most likely agent within the identified agent type will be processed. This module is periodically updated as survey teams provide information, thereby, increasing the accuracy of the prediction.

2.2. METEOROLOGY

A Meteorological (MET) Status Screen can be pulled-down by the incident commander or trainee to review a range of MET conditions to include temperature, relative humidity, wind speed, wind direction. A rapid change in wind direction can adversely affect the staging of hot zone/warm zone operations, location of decon and casualty estimation sites, and access/egress routes. For example, note that warm and cold zone operations need to be upwind of the hot zone area

The manual or automated tracking of meteorological/weather conditions from local weather tracking stations such as airports or news stations is used as input. Information can be automatically downloaded from the Internet or via cell phone connections. Micrometeorological (MICROMET) conditions can be monitored at the incident using remote meteorological stations (sentries) that automatically track environmental conditions and send data to the emergency operations center.

2.3. EMERGENCY DECONTAMINATION AND CASUALTY ESTIMATION

Tracking the process of incident site decontamination and casualty estimation is an overwhelming task for the chemical biological (CB) incident commander. It is a very labor intensive and chaotic process that needs careful managing in order to be successful. In the training mode, the trainee will be challenged with decisions and operations that must be performed in a timely manner in order to be successful. In the decision-aid mode, the system will postulate the probable scenarios based on the potential number of casualties, the kind of agent employed, the dispersion source characteristics and the specific venue where the incident occurred. This will involve logical integration of the Agent Identification (Signs & Symptoms), Hazard Assessment, Clock/Time and Decon/Casualty Estimation processes.

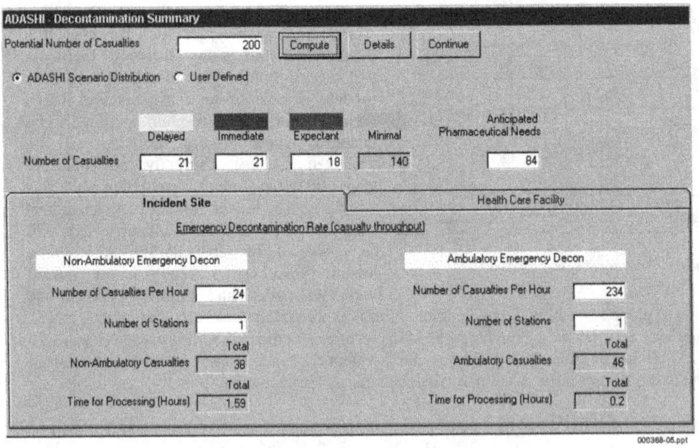

Inputs into this processor will come from pull-down menu icons that will query the trainee or the incident command user on the characteristics of this particular event. Time of incident, MET conditions, venue and hazard source information, and signs and symptoms will be called up from previous entries to verify that these data are still pertinent.

Figure 3. Casualty distribution and applied decontamination time-based effective.

Monitoring weather is very important, especially if temperatures fall below 50 degrees F, due to problems with hypothermia given the amount of water used in a typical decontamination process. This process will provide estimated levels of casualties. This projection of casualties is based upon the START TRIAGE system for casualty estimation. A dynamic tracking tool as a pull down option for the trainee or IC is provided to monitor this multifaceted procedure (Figure 3).

Outputs from the above processor will include emergency decon station (EDS) casualty throughputs (EDS rate) and will monitor both ambulatory and non-ambulatory EDS. ADASHI will assist the trainee or the IC in determining whether decon or casualty estimation rates are adequate; or suggest establishing additional EDS corridors with assistance from existing local support agreements.

2.4. TIME/CLOCK MONITORING AND TRACKING

An integrated Time Monitoring/Tracking module is a key element in the architecture for ADASHI. This algorithm will be linked to all processes that are time-dependent, or require time tracking of some sort.

Incident tracking for the EOC, or the IC, will include actual time monitoring or Real Time, the estimated time the incident occurred, and the elapsed time since the incident occurred. Time tracking will also be embedded into the Personal Protective Equipment (PPE) algorithm. This will allow emergency responders to coordinate proactive personnel accountability process that will track entry/ exit times in the hot and warm zones. The PPE module will also monitor over time what activities these responders are performing and track their potential exposure to environmental hazards based on the particular protective ensemble that is worn.

Many table-top or field training exercises develop an assessment or profile as to how well emergency responders are making decisions; and how well they are actively implementing the training they have taken in this new emergency area. What is not assessed well in the aforementioned training sessions is an active time-dependent measure of their performance, both at the tactical and the incident command levels.

2.5 HAZARD MITIGATION PROCESS

In many cases, actual or potential threats using CB dispersion devices can be mitigated with an effective hazard mitigation reduction protocol. This will require both the explosive ordnance technician (bomb tech) and the hazardous material technician to work hand in hand as an integrated team to reduce the probable airborne hazards from a variety of hazard sources. These sources (or devices) can be as simple as a toxic spill similar to what occurred in Tokyo with the Aum Shinrykyo cult or much more complex with very efficient dispersion systems.

This module will include practical information on how to approach suspect devices, how to mitigate their dispersion effects and how to suppress and contain the hazards to preclude long-term airborne exposure. In many cases these techniques are already used at the technician level by emergency responders across the country. It will also present to the trainee or the emergency responder the logic behind how hazard sources are generated, which sources are of concern, and what venues are more susceptible than others are. After the source characteristics are understood, the deployment of hazard mitigation and containment techniques will be described and situations presented where utilization of these systems makes sense. A trainee will be queried on what alternatives could be used in particular hazard settings to verify the individual's competency level.

2.6. CASUALTY ESTIMATION, TRANSPORT AND MEDICAL TREATMENT PROCESS

The Casualty Estimation, Transport and Medical Treatment process is an integrated sequel to the Decon and casualty estimation process. These processes are intimately related in a CB incident. The Casualty Estimation, Transport and Treatment component expands the knowledge of the trainee in the training mode by prompting and presenting in a time-dependent sequence the myriad roles and responsibilities of the at-site and local health care provider. These responsibilities, which are critically time-dependent include:

— Emergency decontamination and casualty estimation actions at the local clinics and hospitals. The supporting algorithm will receive/transmit two-way feedback between the Signs and Symptoms Algorithm. This will allow for enhanced assessment of the released hazardous agent and how to provide more effective treatment.

   – Post-incident actions will include the performance of medical services as necessary, both at the site and at local hospitals and clinics. Other prompts will direct the user to recommended casualty estimation procedures and emergency decontamination for the self-referred or incident-transported casualties reach the treatment area. It will assist in the set-up of coordination measures for contacting additional medical personnel to bolster the response to a CB mass casualty incident.

3. Training Overlay

Automated multifunction tracking and monitoring can be used as an effective interactive training tool. A trainee can structure a hazardous incident scenario through his inputs on the GUI. The scenario presented then requires specific actions to mitigate the effects of the incident.

ADASHI's processors can then determine the scope of tactical alternatives available and query the trainee using direct questions, memory prompts, etc to help in making an informed decision (Figure 4).

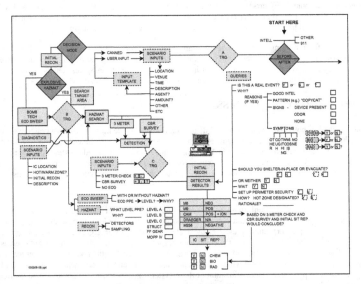

Figure **4**. ADASHI learning concepts.

The Training component can serve as an adjunct to didactic and practical application training to test the concepts and methodologies the trainee has learned. The trainees can enhance their mastery of the various tactical issues in responding to these CBR incidents by working through a bank of queries set up to test the capabilities of these responders. This programmed learning sequence may use "canned" scenarios or the student may opt to create his own. The learning mode will focus on optimizing situational awareness and provide queries and/or guidelines as to what actions might be taken dependent upon the circumstances. The system can be set up to work in a graded or pass-fail mode.

The system can be set up to track data during an event allowing for manual as well as automatic updating of the situation. Prompting and querying will automatically require prompt response by trainees, thereby enhancing their real-time decision-making prowess. Employing fuzzy logic processing, ADASHI interacts with the student in moving toward a general increase in competence and understanding. If the student is very weak, based on the level of monitored improvement, the system will bring up fundamental concepts as a programmed learning tutorial. If the student has displayed more competence, it will "nudge" the student along toward a general level of competency (Figure 5).

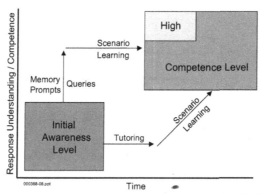

Figure5. ADASHI learning concepts.

This training component will be an integrated and consistent system using directly the operational decision-aid mode processing of this system. All active processing algorithms are identical to the operational mode so that a real-time, realistic scenario is presented to the trainee in the training mode.

Training decisions and prompts are embedded into a flow-processing matrix. The direct expert logic used in the decision-aid mode is employed so the trainee can gain experience, and then can use the system as a support tool for actual operational venues or incidents. The training overlay can be driven as if it were accepting decision inputs identical to those used in the operational mode.

Other templates can be constructed in this training mode to accommodate a variety of operational users to include – US Army Tech Escort Unit, National Guard Raid Teams, United States Marine Corps (USMC) CBIRF, HAZMAT responders, EMS responders, Explosive Ordinance Disposal (EOD) bomb technicians and hospital providers.

Also in this mode, individuals at home can use ADASHI or it can be used in operational team training settings. This training can be accomplished using a distributed learning network or over the Internet. It will be an interactive system coaxing, monitoring and reminding the trainees of the particular situation they are in. As aforementioned this will be accomplished using a comparison algorithm to track trainee responses and compare them with expert situational alternative.

156

4. Bibliography

Journal articles

1. American Medical Association. (1997) Biological Warfare, A Historical Perspective, *Journal of the American Medical Association* **278**, 5.

2. American Medical Association. (1997) Clinical Recognition and Management of Patients Exposed to Biological Warfare Agents, *Journal of the American Medical Association*, **278**, 5.

Book references

1. Canadian Security Intelligence Service. (1995) *Chemical and Biological Terrorism: The Threat According to the Open Literature.*

2. McIntire Peters, K. (2000) *Defending the U.S. Early Bird.*

3. U.S. Army Medical Research Institute of Chemical Defense. (1995) *Medical Management of Chemical Casualties Handbook.*

4. U.S. Army Medical Research Institute of Infectious Diseases. (1998) *Medical Management of Biological Casualties Handbook*, Appendix D, Fort Detrick, Federick, Maryland.

5. U.S. Army Medical Research Institute of Infectious Diseases. (1998) *Medical Management of Biological Casualties.*

6. U.S. Department of Army. (1990) *FM 3-9, Potential Military Chemical/Biological Agents and Compounds.*

7. U.S. Department of Army. (1992) *FM 3-3, Chemical and Biological Contamination Avoidance*, Headquarters, Department of the Army.

8. U.S. Department of the Army. (1993) *FM 3-5 NBC Decontamination.*

9. U.S. Department of the Army. (1986) *1986 procedures for self, equipment, and mass casualties. FM 3-6, Field Behavior of NBC Agents*, Washington, DC.

10. U.S. Department of Transportation. (1996) *1996 North American Emergency Response Guidebook.*

11. U.S. Senate Subcommittee on Investigations. (1995) *Hearing on Global Proliferation of WMD.*